Mathematics and Visualization

Series Editors

Gerald Farin
Hans-Christian Hege
David Hoffman
Christopher R. Johnson
Konrad Polthier
Martin Rumpf

Helwig Hauser
Hans Hagen
Holger Theisel

Editors

Topology-based Methods in Visualization

With 157 Figures, 54 in Color

Springer

Helwig Hauser
University of Bergen
Department of Informatics
Postbok 7803
N-5020 Bergen, Norway
E-mail: Helwig.Hauser@UiB.no

Hans Hagen
Fachbereich Informatik
Technische Universität Kaiserslautern
Postfach 3049
67653 Kaiserslautern, Germany
E-mail: hagen@informatik.uni-kl.de

Holger Theisel
Universität Bielefeld
Technische Fakultät
Postfach 100131
33501 Bielefeld, Germany
E-mail: theisel@techfak.uni-bielefeld.de

ISBN 978-3-642-08977-0 e-ISBN 978-3-540-70823-0

Mathematics Subject Classification: 34-06, 34C05, 34C07, 34C23, 34C30, 34C37, 34C60, 37M99, 54H20, 65-06, 65D05, 65D10, 65D18, 68U20

Springer is a part of Springer Science+Business Media
springer.com

Cover design: *design & production* GmbH, Heidelberg

Preface

Enabling insight into large and complex datasets is a prevalent theme in current visualization research for which different approaches are pursued. Topology-based methods are built on the idea of abstracting characteristic structures such as the topological skeleton from the data and to construct the visualization accordingly. Even though the roots of related research date back to the late 1980's and thereafter, there are currently new demands for and renewed interest in topology-based visualization solutions. Practitioners see the potential for highly efficient and meaningful visualization techniques which communicate a lot more (information, sematics) than just data.

To enable additional stimulation to the again sprouting field of research on topology-based methods in visualization, we initiated an international workshop of about forty researchers from about a dozen European and American places in the beautiful castle of Budmerice in Slovakia. Our intention was to bring together active researchers in a focused two-day event (*TopoInVis*, Sept. 29th & 30th, 2005) to jointly assess the state of the art, to identify trends and joint interests, to discuss and exchange new thoughts on not yet resolved challenges, and to raise new questions and stimulate future research. In a diversified program of 14 presentations, 4 panel discussions, and social activities, we tied in the great tradition of *Dagstuhl* seminars with an atmosphere of informal discussions, open exchange, and personal interaction.

A number of interesting research questions have been addressed throughout the workshop, including intense discussions about topology-based approaches to time-dependent data, an assessment of similarities of as well as differences between topology-based and feature-based approaches, the relation between data topology and semantics in the data, and the question of applicability and fields of potential application (also in contrast to considerations with respect to theory-centric questions).

Organization of this Book

In this book, we present 13 peer-reviewed papers as written results from the workshop. We start with a state-of-the-art report by Laramee et al., followed by a paper about topology-guided visualization of flow simulation data (Peikert & Sadlo). The following papers are about utilizing the scale-space for tracking critical points (Klein & Ertl) and about feature flow fields in an out-of-core setting (Weinkauf et al.). Next, a paper about streamline predicates (Salzbrunn & Scheuermann) is included, right before a comparison of topology-based and feature-based approaches (Hauser et al.), and a paper about superposition effects in topology-based approaches (Ebling et al.). Theisel et al. reflect on the general applicability of topology-based methods. An investigation of swirl and tumble motion in CFD data is described by Garth et al. whereas the application case of advanced design engineering is presented by Trenker. With the paper about a topology-based approach to scalar fields (Bremer & Pascucci) the focus of the book is widened, then including a paper about the role of topology in focus+context visualization (Viola & Gröller) and a paper on reconstruction based on topological changes (Tóth et al.). We think that this book is an interesting snapshot of selected hot topics in topology-based methods in visualization.

Acknowledgements

TopoInVis 2005 has been organized and supported by the VRVis Research Center in Vienna, Austria, the Max Planck Institut für Informatik (MPII) in Saarbrücken and the DFG International Research Training Group "Visualization of Large and Unstructured Datasets" at the University of Kaiserslautern, both in Germany, and the Comenius University in Bratislava, Slovakia.

Conference Room

Budmerice Castle

It is our pleasure to thank all contributors to and all supporters of the workshop. In particular, we wish to extend our sincere gratitude to Andrej Ferko for his manifold support, esp. with the organization of the workshop venue, Christoph Garth for his great help, esp. with the publication process, Robert S. Laramee for helping with web pages and travel management, and Matej Novotný for his manifold assistance, as well as Helmut Doleisch, Georg Rothwangl, Sylvia Kiss, Karin Kraft, and Agnes Knor for lots of help with the organization in Vienna, Marek Zimanyi, Vlado Roth, and Zsolt Tóth for lots of help with the organization in Bratislava and Budmerice, and the staff of Budmerice castle for being really great hosts.

Last but not least, we wish to thank the sponsors of this workshop, including the four organizing institutions as well as the German Research Foundation (DFG), the European Association for Computer Graphics (EUROGRAPHICS) and the Austrian Computer Society (OCG).

Satisfied by many valuable impressions from *TopoInVis 2005* we are looking forward to TopoInVis 2007 to be held in Grimma near Leipzip in spring time.

Wien, Kaiserslautern and Bielefeld,
February 2007

Helwig Hauser
Hans Hagen
Holger Theisel

Group Photo

Contents

Topology-Based Flow Visualization, The State of the Art

Robert S. Laramee[1], Helwig Hauser[2], Lingxiao Zhao[3], and Frits H. Post[4]

[1] Department of Computer Science, Swansea University, UK
r.s.laramee@swansea.ac.uk
[2] VRVis Research Center, Vienna, Austria
Hauser@VRVis.at
[3] Data Visualization Group, Delft University of Technology
{F.H.Post,zlx}@ewi.tudelft.nl

Summary. Flow visualization research has made rapid advances in recent years, especially in the area of topology-based flow visualization. The ever increasing size of scientific data sets favors algorithms that are capable of extracting important subsets of the data, leaving the scientist with a more manageable representation that may be visualized interactively. Extracting the topology of a flow achieves the goal of obtaining a compact representation of a vector or tensor field while simultaneously retaining its most important features. We present the state of the art in topology-based flow visualization techniques. We outline numerous topology-based algorithms categorized according to the type and dimensionality of data on which they operate and according to the goal-oriented nature of each method. Topology tracking algorithms are also discussed. The result serves as a useful introduction and overview to research literature concerned with the study of topology-based flow visualization.

1 Introduction

Research in topology-based flow visualization is making rapid advances. Helman and Hesselink introduced the visualization community to the notion of flow topology in 1989 [21, 23]. Classical flow oriented topology research is based on the detection and classification of critical points in the vector field, as shown in Figure 2. What makes topology-based methods attractive is their ability to represent very large data sets in a concise and compact manner. Unlike other flow visualization approaches (Figure 1), critical points of a data set are extracted and the relationships between those points are depicted accordingly. We refer the reader to Abraham and Shaw for an introduction to topological analysis [1].

Topology-based research in flow visualization has come a long way since 1989 –the progress of which we will describe in Section 2. Yet, despite the many

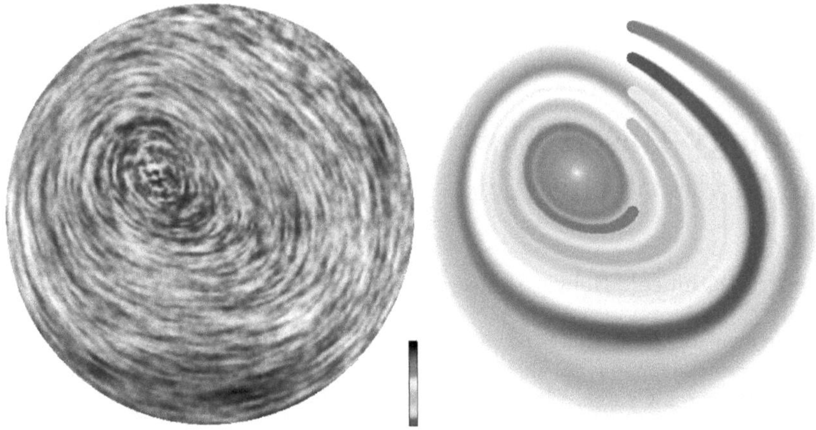

Fig. 1. Visualization of flow around a critical point using texture advection and dye injection [35]. In contrast to these methods, topology-based methods extract and visualize critical points directly (colorplate on p. 200).

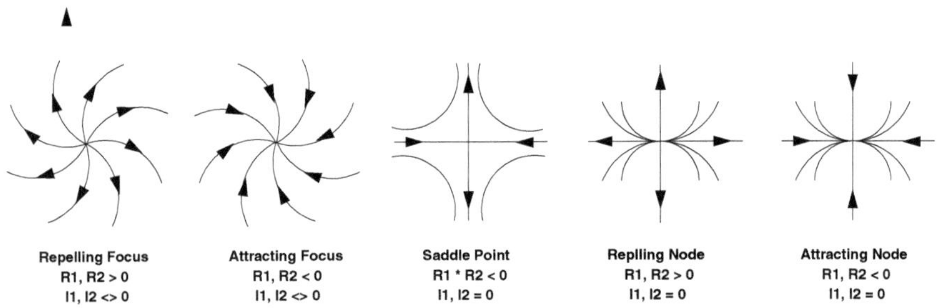

Fig. 2. Vector field topology: critical points are usually classified by the eigenvalues of the Jacobian [21]. R represents the real components and I the imaginary components of the Jacobian.

advances, there are still many unanswered questions in the field of topology-based research. There are still topic areas completely untouched by researchers at the time of this writing, e.g., vector and tensor field topology simplification in three-dimensions, for both steady and time-dependent (or unsteady) data.

Here, we summarize the progress that has been made up to this point in the field. We introduce a novel classification of topology-based methods in flow visualization based on topology extraction and simplification of vector and tensor fields (Section 2). The classification points out clearly those areas rich in previous work and some areas which still remained unaddressed by the visualization community.

2 Topology-Based Methods in Flow Visualization, The State of the Art

In this section, we review the current state of the art in topology-based methods in flow visualization. We start off with a description of our classification before describing the algorithms themselves. Our overview relates different research results with one another and highlights relative advantages and disadvantages of each approach.

2.1 Classification

		Topology Extraction			Topology Simplification	
Scalar		Static	Dynamic		Static	Dynamic
Data	2D	[19]		2D	[5] [6]	
	2.5D	[44] [26] [45]	[7]	2.5D	[13] [16] [61]	
	3D	[40] [62]		3D	[17]	
Vector		**Steady**	**Unsteady**		Steady	Unsteady
Field	**2D**	[55] [56] [87] [73]	[21] [53] [80] [89] [83] [68] [71] [72]	2D	[8] [10] [9] [38] [77] [82] [76] [64] [67] [66]	
Data	**2.5D**	[22] [30] [32] [75] [84]		2.5D		
	3D	[37]$_v$ [27]$_v$ [59]$_v$ [29]$_v$ [31]$_v$ [51]$_v$ [18]$_v$ [46]$_v$ [28]$_v$ [41]$_v$ [58]$_v$ [54]$_v$ [23] [39] [88] [43] [42] [70] [41] [14] [74] [60] [86] [33]	[2]$_v$ [3]$_v$ [49]$_v$ [4]$_v$ [15]	3D	[85]	
Tensor		Steady	Unsteady		Steady	Unsteady
Field	2D	[91]	[11] [79]	2D	[78] [82] [81]	
Data	2.5D			2.5D		
	3D	[24] [90] [92] [93] [25]		3D		

Table 1. An overview and classification of topology-based methods in visualization. Research is divided up into topology extraction and topology simplification literature. Methodology is further classified according to scalar vs. vector vs. tensor field data analysis. Finally, a sub-classification is made based on data dimensionality, both spatial and temporal. References are listed in chronological order within each spatio-temporal dimensionality. In Section 2. we focus on the research with **bold** emphasis–topological analysis of vector field data. References subscripted with a $_v$ denote research related to vortex core extraction.

Table 1 illustrates our classification of topology-based methods in visualization. At the broadest level of classification, we have divided up the literature into work that focuses on either *extraction* of topological features, i.e., topological analysis or *simplification* of a given topology. Conceptually, simplification can be thought of as an extension of extraction. We separate the literature focused on simplification because much of it is dedicated to simplification of an *a priori* topology, especially in the area of flow visualization–the focus of this overview. We have further divided up the literature into *vector* and *tensor* field analysis. Each sub-classification is then further classified based on the spatial and temporal dimensionality of the vector or tensor field data to which the respective algorithm is applied. The topology research on scalar data is divided into *static* and *dynamic* cases rather than steady and unsteady in order to be more general. Dynamic analysis of scalar data sets can also include a transformation from one static surface to another surface [7]. Within a single spatio-temporal dimension, references are listed in chronological order. Our overview focuses on those categories with **bold** emphasis, namely, topological analysis and extraction of vector field data. The focus on vector field analysis was chosen in order to limit the scope of the review. The topics of scalar and tensor field topology can be covered in future state of the art reviews. Note that within the category of **3D, vector field extraction**, literature which focuses on vortex core extraction is denoted with subscript ($_v$). We now describe the literature in increasing order of dimensionality, grouped together by topic. Another overview is given by Scheuermann and Tricoche [57].

Although the topology of *scalar fields* serves as a third category of research, our review of the literature does not focus on the topological analysis of scalar fields [19, 26, 62] which includes the extraction of features such as ridge and valley lines and extremal features. Our survey of scalar topology analysis is also not exhaustive, but supplies the reader references for further reading. Here we briefly mention some research in the field. Monga et al. [44] compute ridge lines on isointensity surfaces in 3D volume data and use them for data registration and automatic atlas generation. Interrante et al. [26] use ridge and valley lines in order to perceptually, enhance the visualization of multiple, transparent surfaces in 3D. Szymczak and Vanderlyde describe an algorithm that extracts topologically simple isosurfaces [61]. Morse theory has been applied to extract the topology of arbitrary surfaces by Ni et al. [45].

2.2 Topology Extraction of Vector Field Data

2D, Steady

Extraction of Higher-Order Critical Points: Most critical point detection algorithms are based on piecewise linear or bilinear approximation. These methods do not properly represent local topology if nonlinear behavior is present. Scheuermann et al. [55, 56] choose a polynomial approximation in areas with nonlinear behavior and apply a suitable visualization–streamlines seeded at the critical points with additional annotations.

Extraction of Closed Streamlines: Wischgoll and Scheuermann [87] present an algorithm for detecting closed streamlines in planar flows. Closed streamlines are of interest because they may indicate regions of recirculating flow. It is based on monitoring streamlines as they enter, exit, and re-enter cells of the vector field domain. We urge the reader to use caution when interpreting the visualization results. This is because an spatial dimension inherent to the applied domain has been left out of the analysis.

The first approach to detecting closed streamlines in planar flow was based on monitoring polygon-based entrance and exit events of a streamline during integration [87], This approach is extended to time-dependent flows by Wischgoll et al. [89]. At each time step, closed streamlines are extracted. Afterwards, a time-dependent correspondence between individual streamlines is computed. Theisel et al. [73] present an alternative approach to computing closed streamlines. A 2D vector field is transformed into a 3D vector field. This can be done by representing time as a third spatial dimension. Then streamsurfaces are seeded in the 3D domain. Finally, closed streamlines are detected by intersecting streamsurfaces. The difference to previous work is that this approach avoids mesh-based dependency, e.g., examining and testing individual mesh polygons.

Vector Field Design: Theisel presents a novel method that allows the user to design higher order vector fields of arbitrary topology [64]. The technique is based on control polygons that let the user specify the characteristics of critical points. This enables a mechanism by which to test topology extraction algorithms. The result can also be used for compression purposes. We note that this research does not fit cleanly into our classification partially because it spans more than one area.

2D, Unsteady

Detection and Classification of Critical Points: Helman and Hesselink introduced the visualization community to flow topology [21]. Their analysis included the detection, classification, and visualization of critical points in planar flows (Figure 3). They applied their algorithms to both steady-state and unsteady flow. They represent time as a third spatial dimension for the case of time-dependent, planar flow.

Vortex Detection Based on Streamline Geometry: Sadarjoen and Post [53] present two methods for detecting vortex structures in 2D vector fields. They are both based on an analysis of streamline geometry. The first method uses local cumulations of curvature that may indicate a group of vortices in very close proximity to one another. The second method looks at the curvature of a single streamline and computes a winding angle–a metric of geometric curvature. One advantage of this technique is that it detects weak vortices because it does not depend on velocity magnitude at a single point. A disadvantage, however, is the large number of streamlines that must be seeded and computed in order to maintain complete coverage of the flow.

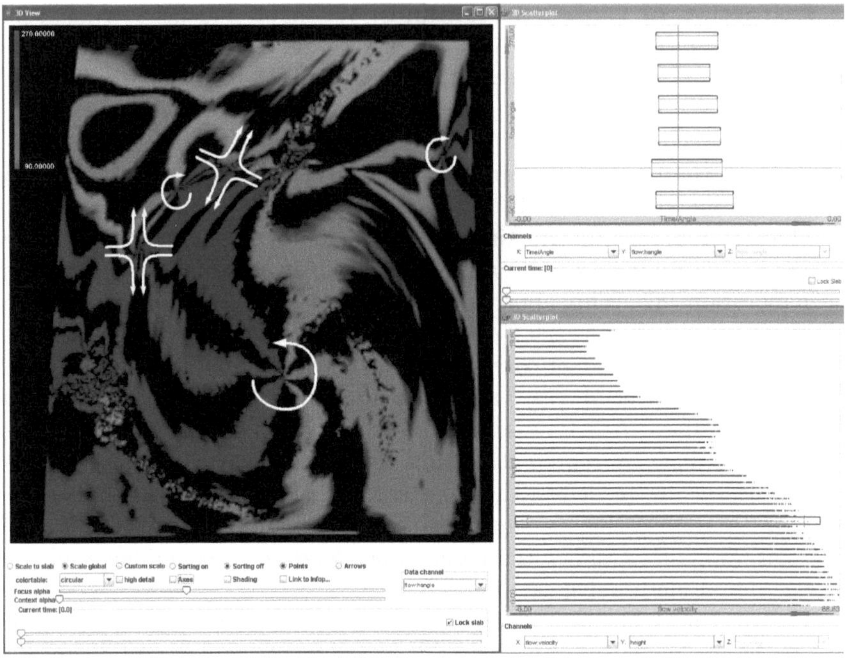

Fig. 3. Here, the topology of a hurricane is visualized and depicted with the SimVis system [12] (colorplate on p. 200).

Detection of Topological Transitions: A novel topology-based method for the visualization of time-dependent 2D flows is given by Tricoche et al. [80]. Extending the work of Helman and Hesselink [21, 23], they identify and visualize topological transitions–the qualitative change of topology structure from one stable state to another over time. Three types of transitions are investigated: (1) a Hopf-like transition–a transition of a singular point from an attracting focus (i.e. sink) to a repelling focus (i.e. a source), (2) a fold-like transition–the pairwise annihilation or creation of a saddle and a source or sink, (3) a basin transition–the case when two saddle points start independent of one another, join briefly, and again separate. Again we caution the reader when interpreting these results. A spatial dimension inherent to the original domain has been omitted from the analysis.

Critical Point Tracking: Theisel and Seidel introduce an alternative critical point tracking method for 2D, unsteady flow based on streamlines [68]. The temporal dimension of the planar flow is represented as a third spatial dimension and streamlines are traced along critical points as they evolve. This space-time representation is called a feature flow field. In addition to visualizing the path of critical points over time, events such as fold bifurcations are visualized.

Streamline and Pathline Oriented Topology: Topological methods often segment vector fields using curves based on streamlines, e.g., separatrices or streamsurfaces such as separating streamsurfaces. In addition to streamline oriented topology, Theisel et al. [71, 72] also consider pathline oriented topology. In the study of streamline oriented topology, they propose new approaches to detect bifurcations like saddle connections and cyclic fold bifurcations. Saddle connections are bifurcations that appear when two separatrices originating from saddle points coincide. A cyclic fold bifurcation is the case of when two closed streamlines collapse and disappear. The also propose a novel approach to detect and track closed streamlines in 2D, time-dependent vector fields. In the study of pathline oriented topology, they segment the vector field into regions where pathlines show attracting, repelling, or saddle-like behavior.

Vector Field Comparison: Although it does not fit cleanly into our classification, we briefly mention a closely related topic–vector field comparison. Theisel et al. [65] introduce a topology-based metric by which vector fields can be compared or related to one another. Preliminary approaches based on comparison metrics (i.e., distance measures) were based on local deviations of direction and magnitude of flow vectors [20, 63]. These previous distance functions yield a fast comparison of vector fields, but do not take into account any structural information. Levin et al. [36] introduce the first topology-based approach to vector field metrics with the Earth Mover's Distance (EMD [52]), a technique from image retrieval. The limitations of this algorithm are that: (1) it's critical point coupling strategy does not consider the location of critical points in the vector fields and (2) all critical points are compared to one another which can lead to a worst case complexity of $o(n!)$ where n is the number of critical points. To overcome these critical point coupling limitations, Theisel et al. [65] introduce a comparison metric that uses feature flow fields [68].

2.5D, Steady

Separation and Attachment Lines: Separation and attachment lines correspond to loci where flow leaves or converges at a surface. Prior to Kenwright [30], the only algorithm that could automatically detect separation and attachment lines was presented by Helman and Hesselink [22]. Previous approaches were generally based on observations. Helman and Hesselink's technique is based on vector field topology. Their algorithm detects closed separation lines, that is, lines that begin at a saddle or node and end at another saddle or node. Kenwright's algorithm also detects open separation, i.e., lines that do not always start or end at critical points in the vector field. This algorithm is based on phase plane analysis.

Kenwright et al. [32] expand the work of Kenwright [30] by introducing another algorithm, the parallel vector algorithm, for detecting open separation and attachment lines. The parallel vectors algorithm is based on the observation that one of the eigenvector directions was always parallel to the local streamlines in regions where streamlines asymptotically converged. The advantage of this approach is that it provides a local test that may be performed at any point in the vector field. Kenwright et al. show that the parallel vectors algorithm is slightly superior to their previous algorithm (called the phase plane algorithm), however, it is more difficult to implement. The phase plane algorithm uses self-contained analysis within each triangle, making it well suited for unstructured meshes. The parallel vector algorithm requires calculation of vector gradients on irregular triangulations. But for curvilinear meshes, the parallel vector algorithm is best because vector gradients can be calculated using central differences. The parallel vector algorithm also resolves the line discontinuity problem associated with the phase plane algorithm.

Tricoche et al. [75] propose a method for the detection of separation of attachment lines in 2D flows defined over arbitrary surfaces in 3D. They build primarily on the work of Kenwright and Haimes [30, 32] by improving performance. They do so by using both local flow properties and global structural information such that feature searching and extraction is fast and accurate.

Boundary Switch Connectors: Weinkauf et al. [84] extend the work of Theisel et al. [70] with the introduction of *boundary switch connectors*, a topological element that complements saddle connectors. Theisel et al. [70] considered separation surfaces emanating from saddle points only. Weinkauf et al. [84] extend this work to include separating surfaces starting from boundary switch curves. The intersection of separating surfaces emanating from boundary switch curves results in boundary switch connectors.

3D, Steady

Vortex Core Line Extraction: Sujudi and Haimes [59] present a line-based vortex core extraction algorithm that locates points that satisfy the following two criteria: (1) the velocity gradient tensor contains complex eigenvalues and (2) the velocity in the plane perpendicular to the real eigenvector is zero. The individual points are then connected to form the vortex core line. The disadvantage here is that it is not always possible to form a continuous line. This problem is addressed by Haimes and Kenwright [18] who present adapt the algorithm to be face-based rather than cell-based.

Vortices can cause many undesirable effects for aircraft, such as reduced lift and noise. They can lead to structural fatigue and even premature airframe failure in severe cases. Kenwright and Haimes [29, 31] applied the eigenvector method of Sujudi and Haimes to flow analysis around an aircraft.

Roth and Peikert build on the work of Sujudi and Haimes [59] by introducing a higher-order method for vortex core line extraction. While the eigenvector method of Sujudi and Haimes [59] is correct for linear vector fields, it fails to detect curved vortex core lines, especially in the case of turbomachinery data sets. Roth and Peikert demonstrated this limitation previously [50]. Their method overcomes the previous limitations stemming from the use of a linear vector field for vortex core line extraction by introducing higher-order derivatives that can be used to detect bent vortex cores.

This vortex core line extraction algorithm is later formulated at a higher level of abstraction, namely as a parallel vectors operator by Peikert and Roth [46]. The basic idea behind the parallel vectors approach is to derive two vector fields from a given 3D vector field such that vortex core lines are locations where the two derived vector fields are parallel.

Some vortex core extraction methods, like that from Jeong and Hussain [27], can be described as Galilean invariant, i.e., they are invariant when a constant vector field is added. This is because their computation uses only derivatives of the vector field. Many vortex core line extraction algorithms are Galilean variant because they depend on a certain reference [2, 3, 46, 59]. Sahner et al. [54] present an approach to extracting vortex core lines that is Galilean invariant, i.e., the result does not depend on the frame of reference. The extracted features remain unchanged when adding a constant vector field. They do so by considering ridge or valley lines of Galilean invariant vortex region quantities.

Vortex Core Region Extraction: A general problem with vortex core line extraction algorithms is their computational complexity and that they may generate more than one vortex core line within a vortex core region. Mahrous et al. [41] present a vortex core region detection based on Sperner's lemma–adapting a notion from combinatorial topology. The approach analyzes the behavior of a vector field based on the vectors found at the boundaries of each grid cell. Velocity vectors exhibit characteristic patterns in the neighborhood of a vortex. The algorithm searchers for these patterns.

In our overview, we focus on vortex core line extraction rather than vortex core region extraction. Thus the method of Jeong and Hussain, known as the λ_2 method [27] is not described in detail here (Stegmaier and Ertl present a GPU-based implementation of the λ_2 method [58]). Similarly, we do not focus on vortex core extraction based on isosurface extraction in a scalar field [37]. A more general overview of vortex analysis from a feature-based flow visualization point of view is given by Post et al. [47].

Separating Surfaces: Helman and Hesselink build on their previous work [21] and extract surface topology and separating surfaces of flow in 3D [23]. A surface topology skeleton is extracted and visualized by projecting the 3D vector field in the neighborhood of the surface onto the plane tangent to the body and applying a 2D detection algorithm. They also compute streamsurfaces which separate 3D vector fields into disparate regions of flow. Included is a description of how these streamsurfaces are tessellated in an efficient manner. They also uses icons such as arrows and disks to display critical points in 3D.

Mahrous et al. [41, 42] present an algorithm for efficient computation of separatrices in 3D vector fields. They present methods that accelerate the extraction of separatrices. Enhancements are made to reduce the number of sample streamlines and their length. Streamlines are seeded in a more meaningful

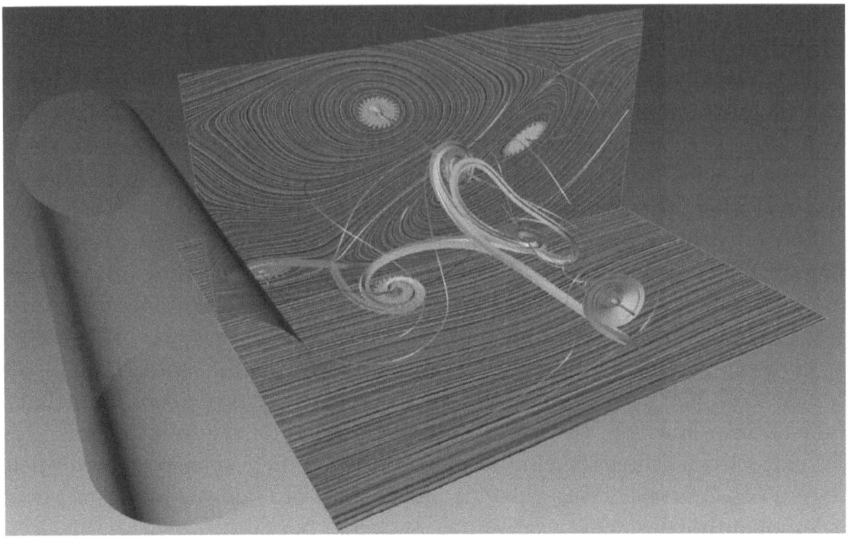

Fig. 4. Visualization of flow past a circular cylinder using critical points and saddle connectors [70] (colorplate on p. 201). *Image courtesy of H. Theisel et al.*

and a efficient matter rather than using a brute-force approach of seeding streamlines at all cell locations. Texture advection is applied to stream surfaces by Laramee et al [34].

Dynamical Systems: Löffelmann and Gröller [39] visualize the topology of dynamical systems. Dynamical systems provide a mathematical model comprised of a set of state variables whose goal is to characterize real world phenomena, e.g., a stock market, a chemical reaction, or a food chain. Their visualization couples characteristic streamlines emanating from fixed points in the domain with a thread of streamlets. The characteristic streamlines play the role of seed points for a thread of streamlets. The large number of streamlets provide more information about the behavior of the dynamical system in the neighborhood its characteristic trajectories. Thus a trade-off between domain coverage and perceptibility is realized in 3D.

Detection of Closed Streamlines: Wischgoll and Scheuermann [88] extend their previous work [87] of detecting closed streamlines to 3D vector fields. The algorithm is based on preventing infinite cycling during streamline integration. **Saddle Connectors:** Theisel et al. [70] introduce a new topological element of vector fields called a *saddle connector*. A saddle connector is a streamline that joins two saddle points in a vector field (Figure 4). A saddle connector is found essentially by computing the intersection of the separation surfaces of two saddle points. These topological structures achieve a visually sparser, more compact topological representation of the vector field, thus

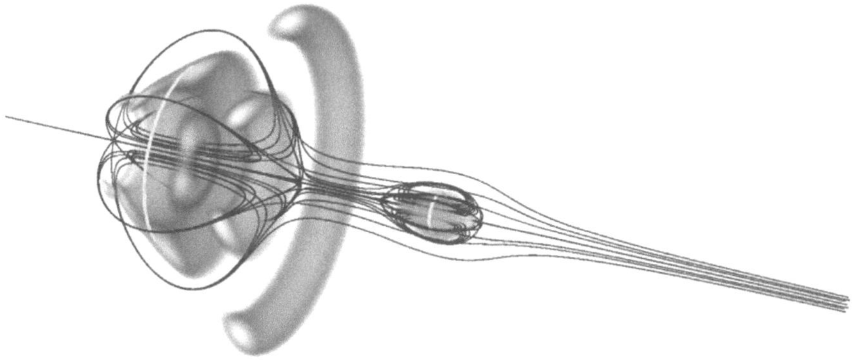

Fig. 5. The visualization of a vortex breakdown bubble. Flow topology is depicted with stagnation points in red, singularity paths in yellow and streamlines in blue [74] (colorplate on p. 201). *Image courtesy of X. Tricoche et al.*

avoiding the visual complexity associated with showing too many separating streamsurfaces.

Hybrid Visualization and Vortex Breakdown: Tricoche et al. [74] use a combination of 3D volume rendering of a vector field's scalar fields with vector field topology projected onto a moving cutting plane. The goal is to gain insight into the behavior of vortex breakdowns with this novel hybrid visualization (Figure 5).

Critical Point Modeling and Classification: Weinkauf et al. [85] extend the work of Theisel [64] for designing vector fields. In particular they: (1) model 3D vector fields of arbitrary topology. Previously, only first order points and the index of higher order critical points were considered [43], (2) introduce a complete classification of 3D critical points and (3) adapt the notion of saddle connectors in order to model the intersection curves of separation surfaces. Thus, the problem of modeling a vector field is reduced to the problem of modeling the topological skeleton using control polygons.

Weinkauf et al. [86] extend the work of Tricoche et al. [77] to 3D. They introduce an extraction and classification scheme for higher order critical points in 3D. The approach is based on enclosing a critical point, or a cluster of critical points by a bounding surface. The properties of the vector field at the boundary surface are then examined in detail, i.e., subsets of the surface are divided up into inflow, outflow, hyperbolic, and elliptic regions of flow. The classification of critical points in 3D is then determined by the corresponding regions on the bounding surface. The simplified structure of the flow within the bounded regions is then visualized with an appropriate icon(s).

Applications of Topology-Based Flow Visualization: Sun et al. [60] apply a topological analysis to visualize the power flow through a C-shaped nano-aperture. Such an aperture may be very effective at power transmission with applications including data storage, particle manipulation, and nanoscale photonic devices. Their topological analysis of this data set results in a heightened understanding of the critical factors affecting power transmission of these apertures including: polarization effects, efficiency, the size of interaction regions, resonant transmissions, and more.

Laramee et al. [33] apply topology-based flow visualization methods in order to gain insight into the behavior of flow through a cooling jacket. This application is discussed in more detail in a following chapter. Other applications of topology-based flow visualization are discussed by Garth et al. [15] and Tricoche et al. [74].

3D, Unsteady

Vortex Core Line Extraction and Tracking: Banks and Singer [2, 3] developed an algorithm for vortex tube reconstruction based on the assumption that a vortex core is a vorticity line–a streamline in the vorticity field. and pressure is a minimum in the core. The algorithm consists of four basic steps: (1) compute the vorticity along a vortex core line (seeded based on threshold vorticity magnitude and pressure), (2) predict the next point along the core line by stepping in the vorticity vector's direction, (3) compute the vorticity at the new predicted point, and (4) update (or correct) the point to the location of minimum pressure in the plane perpendicular to the core.

Reinders et al. [49] present an application which detects and tracks vortex tubes in flow past a tapered cylinder. First, they apply the winding-angle method [53] is used to detect the vortices on a number of horizontal slices. Second, the 3D vortex tubes are constructed from the 2D vortices by applying a spatial feature tracking procedure based on attributes of the vortices [48]. The same feature tracking algorithm is then applied in the temporal domain for vortex core tracking.

Theisel et al. [69] describe a novel method to extract parallel vectors [46] based on the use of feature flow fields [68]. They derive appropriate vector fields such that vortex core lines appear as streamlines (in the feature flow fields). Thus, the extraction of vortex core lines is reduced to a well-known streamline integration computation. They also introduce a novel classification of transitions (or events) associated with time-dependent vortex core lines as well as the methodology used in tracking core lines. The classification includes: (1) saddle transitions, (2) closed collapse transitions, (3) and inflow and outflow boundary transitions.

Singularity Tracking and Vortex Breakdown: Garth et al. [15] present a method to efficiently track singularities in 3D, unsteady flow. The method also applies to data defined on unstructured grids. Conceptually, it is an extension of the work of Tricoche et al. [83]. The concept of a singularity index is discussed and extended from the well known 2D case to the more complex 3D domain. The results are particularly insightful for the study of vortex breakdown. Occurrences are vortex breakdown (or bursting) are correlated with local extrema in physical quantities and visualized with corresponding views from information visualization.

2.3 Discussion and Future Prospects

Table 1 clearly illustrates those areas with a heavy concentration of topology-based research, e.g., 3D steady-state, and those areas with little to no work. In fact, Table 1 highlights areas that remain untouched up to this point in time, e.g., topology simplification in 3D tensor fields. Other areas still requiring research work include:

- Interactive techniques to support topology extraction and tracking: At present, topology-based techniques are, in general, still slower relative to traditional flow visualization techniques such as particle tracing or texture-advection methods.
- Extraction and analysis of new types of topological structures: Surely, not all important topological structures have been clearly identified and studied.
- Integration of topology-based methods with other flow visualization techniques such as texture advection: A topological skeleton by itself, sometimes leaves out other important properties of the flow such as downward and upstream direction.
- The practical application of topological methods outside the visualization community: Still, much work remains to be done in the application of topology-based flow visualization to data sets from industry or some application domain area in order to demonstrate their utility in a convincing manner.
- More theoretical development to support cognition of results: Topological analysis still leaves open questions with respect to interpretation of the results. For example, how do we interpret pathline-oriented topology? More theory may be needed to aid such cognition.

Thus, the field of topology-based methods in visualization is still rich in unsolved problems.

However, there may be reasons why so much of the spatio-temporal domain in our classification remains virtually unexplored in the research literature. Reasons may include high levels of complexity and applicability to real-world problem domains. We discuss possible reasons for this in a later chapter.

Acknowledgments

The authors thank all those who have contributed to this research including AVL (`www.avl.com`), the Austrian research program Kplus (`www.kplus.at`). Some CFD simulation data is courtesy of AVL.

References

1. R. H. Abraham and C. D. Shaw. *Dynamics - the Geometry of Behavior.* Addison-Wesley, 1992.
2. D. C. Banks and B. A. Singer. Vortex Tubes in Turbulent Flows: Identification, Representation, Reconstruction. In *Proceedings IEEE Visualization '94*, pages 132–139, October 1994.
3. D. C. Banks and B. A. Singer. A Predictor-Corrector Technique for Visualizing Unsteady Flow. *IEEE Transactions on Visualization and Computer Graphics*, 1(2):151–163, June 1995.
4. D. Bauer and R. Peikert. Vortex Tracking in Scale-Space. In *Proceedings of the Symposium on Data Visualisation 2002*, pages 233–240. Eurographics Association, 2002.
5. P.-T. Bremer, H. Edelsbrunner, B. Hamann, and V. Pascucci. A Multi-resolution Data Structure for Two-Dimensional Morse-Smale Functions. In *Proceedings IEEE Visualization 2003*, pages 139–146, October 2003.
6. P.-T. Bremer, H. Edelsbrunner, B. Hamann, and V. Pascucci. Topological Hierarchy for Functions on Triangulated Surfaces. *IEEE Transactions on Visualization and Computer Graphics*, 10(4):385–396, July/Aug 2004.
7. C. Carner, M. Jin, X. Gu, and H. Qiu. Topology-driven Surface Mappings with Robust Feature Alignment. In *Proceedings IEEE Visualization 2005*, pages 543–550, October 2005.
8. W. de Leeuw and R. van Liere. Visualization of Global Flow Structures Using Multiple Levels of Topology. In *Data Visualization '99 (VisSym '99)*, pages 45–52. May 1999.
9. W. de Leeuw and R. van Liere. Multi-level Topology for Flow Visualization. *Computers and Graphics*, 24(3):325–331, June 2000.
10. W. C. de Leeuw and R. van Liere. Collapsing Flow Topology Using Area Metrics. In *Proceedings IEEE Visualization '99*, pages 349–354, 1999.
11. T. Delmarcelle and L. Hesselink. The Topology of Symmetric, Second-Order Tensor Fields. In *Proceedings IEEE Visualization '94*, 1994.
12. H. Doleisch, M. Mayer, M. Gasser, P. Priesching, and H. Hauser. Interactive Feature Specification for Simulation Data on Time-Varying Grids. In *Conference on Simulation and Visualization 2005 (SimVis 2005)*, pages 291–304, 2005.
13. J. El-Sana and A. Varshney. Controlled Simplification of Genus for Polygonal Models. In *Proceedings IEEE Visualization '97*, pages 403–412, October 1997.
14. C. Garth, X. Tricoche, T. Salzbrunn, T. Bobach, and G. Scheuermann. Surface Techniques for Vortex Visualization. In *Data Visualization, Proceedings of the 6th Joint IEEE TCVG–EUROGRAPHICS Symposium on Visualization (VisSym 2004)*, pages 155–164, May 2004.
15. C. Garth, X. Tricoche, and G. Scheuermann. Tracking of Vector Field Singularities in Unstructured 3D Time-Dependent Datasets. In *Proceedings IEEE Visualization 2004*, pages 329–335, 2004.
16. I. Guskov and Z. Wood. Topological Noise Removal. In *Proceedings, Graphics Interface 2001*, pages 19–26, 2001.
17. A. Gyulassy, V. Natarajan, V. Pascucci, P.-T. Bremer, and B. Hamann. Topology-based Simplification for Feature Extraction from 3D Scalar Fields. In *Proceedings IEEE Visualization 2005*, pages 535–542, 2005.

18. R. Haimes and D. Kenwright. On the Velocity Gradient Tensor and Fluid Feature Extraction. Technical Report AIAA Paper 99–3288, American Institute of Aeronautics and Astronautics, 1999.
19. R. M. Haralick. Ridges and valleys on digital images. *Computer Vision, Graphics, and Image Processing*, 22(1):28–38, April 1983.
20. B. Heckel, G. H. Weber, B. Hamann, and K. I. Joy. Construction of vector field hierarchies. In *Proceedings IEEE Visualization '99*, pages 19–26, 1999.
21. J. L. Helman and L. Hesselink. Representation and Display of Vector Field Topology in Fluid Flow Data Sets. *IEEE Computer*, 22(8):27–36, August 1989.
22. J. L. Helman and L. Hesselink. Surface Representations of Two- and Three-Dimensional Fluid Flow Topology. In *Proceedings IEEE Visualization '90*, pages 6–13, 1990.
23. J. L. Helman and L. Hesselink. Visualizing Vector Field Topology in Fluid Flows. *IEEE Computer Graphics and Applications*, 11(3):36–46, May 1991.
24. L. Hesselink, Y. Levy, and Y. Lavin. The Topology of Symmetric, Second-Order 3D Tensor Fields. *IEEE Transactions on Visualization and Computer Graphics*, 3(1):1–11, March 1997.
25. M. Hlawitschka and G. Scheuermann. HOT Lines: Tracking Lines in Higher Order Tensor Fields. In *Proceedings IEEE Visualization 2005*, pages 27–34, 2005.
26. V. Interrante, H. Fuchs, and S. Pizer. Enhancing Transparent Skin Surfaces with Ridge and Valley Lines. In *Proceedings IEEE Visualization '95*, pages 52–59, 1995.
27. J. Jeong and F. Hussain. On the Identification of a Vortex. *Journal of Fluid Mechanics*, 285:69–94, 1995.
28. M. Jiang, R. Machiraju, and D. Thompson. A Novel Approach to Vertex Core Region Detection. In *Proceedings of the Symposium on Data Visualisation 2002 (VisSym '02)*, pages 217–225. Eurographics Association, 2002.
29. D. Kenwright and R. Haimes. Vortex Identification–Applications in Aerodynamics. In *Proceedings IEEE Visualization '97*, pages 413–416, November 1997.
30. D. N. Kenwright. Automatic Detection of Open and Closed Separation and Attachment Lines. In *Proceedings IEEE Visualization '98*, pages 151–158, 1998.
31. D. N. Kenwright and R. Haimes. Automatic Vortex Core Detection. *IEEE Computer Graphics and Applications*, 18(4):70–74, July/August 1998.
32. D. N. Kenwright, C. Henze, and C. Levit. Features Extraction of Separation and Attachment Lines. *IEEE Transactions on Visualization and Computer Graphics*, 5(2):135–144, 1999.
33. R. S. Laramee, C. Garth, H. Doleisch, J. Schneider, H. Hauser, and H. Hagen. Visual Analysis and Exploration of Fluid Flow in a Cooling Jacket. In *Proceedings IEEE Visualization 2005*, pages 623–630, 2005.
34. R. S. Laramee, C. Garth, J. Schneider, and H. Hauser. Texture-Advection on Stream Surfaces: A Novel Hybrid Visualization Applied to CFD Results. In *Data Visualization, The Joint Eurographics-IEEE VGTC Symposium on Visualization (EuroVis 2006)*, pages 155–162,368. Eurographics Association, 2006.
35. R. S. Laramee, D. Weiskopf, J. Schneider, and H. Hauser. Investigating Swirl and Tumble Flow with a Comparison of Visualization Techniques. In *Proceedings IEEE Visualization 2004*, pages 51–58, 2004.
36. Y. Lavin, R. Kumar Batra, and L. Hesselink. Feature comparisons of vector fields using earth mover's distance. In *Proceedings IEEE Visualization '98*, pages 103–110, 1998.

37. Y. Levy, D. Degani, and A. Seginer. Graphical Visualization of Vortical Flows by Means of Helicity. *AIAA Journal*, 28:1347–1352, 1990.
38. S. K. Lodha, J. C. Renteria, and K. M. Roskin. Topology Preserving Compression of 2D Vector Fields. In *Proceedings IEEE Visualization 2000*, pages 343–350, 2000.
39. H. Löffelmann and M. E. Gröller. Enhancing the Visualization of Characteristic Structures in Dynamical Systems. In *Proceedings of the 9th Eurographics Workshop on Visualization in Scientific Computing*, pages 35–46, 1998.
40. K. Ma and V. Interrante. Extracting Feature Lines from 3D Unstructured Grids. In *Proceedings IEEE Visualization '97*, pages 285–292, 1997.
41. K. Mahrous, J. C. Bennett, G. Scheuermann, B. Hamann, and K. I. Joy. Topological segmentation in three-dimensional vector fields. *IEEE Transactions on Visualization and Computer Graphics*, 10(2):198–205, 2004.
42. K. M. Mahrous, J. C. Bennett, B. Hammann, and K. I. Joy. Improving Topological Segmentation of Three-dimensional Vector Fields. In *Proceedings of the Joint Eurographics - IEEE TCVG Symposium on Visualization (VisSym 2003)*, pages 203–212, 2003.
43. S. Mann and A. Rockwood. Computing Singularities of 3D Vector Fields with Geometric Algebra. In *Proceedings of IEEE Visualization 2002*, pages 283–290, October 2002.
44. O. Monga, S. Benayoun, and O. Faugeras. From Partial Derivatives of 3D Density Images to Ridge Lines. In *Proceedings of the IEEE Computer Society Conference on Computer Vision and Pattern*, pages 163–169, 1992.
45. X. Ni, M. Garland, and J. C. Hart. Fair Morse Functions for Extracting the Topological Structure of a Surface Mesh. *ACM Transactions on Graphics (SIGGRAPH 2004)*, 23(3), 2004.
46. R. Peikert and M. Roth. The Parallel Vectors Operator - A Vector Field Visualization Primitive. In *Proceedings of IEEE Visualization '99*, pages 263–270. IEEE Computer Society, 1999.
47. F. H. Post, B. Vrolijk, H. Hauser, R. S. Laramee, and H. Doleisch. The State of the Art in Flow Visualization: Feature Extraction and Tracking. *Computer Graphics Forum*, 22(4):775–792, Dec. 2003.
48. F. Reinders, F. H. Post, and H. J. W. Spoelder. Visualization of Time-Dependent Data with Feature Tracking and Event Detection,. In *The Visual Computer*, volume 17(1), pages 55–71. 2001.
49. F. Reinders, I. A. Sadarjoen, B. Vrolijk, and F. H. Post. Vortex Tracking and Visualisation in a Flow Past a Tapered Cylinder. In *Computer Graphics Forum*, volume 21(4), pages 675–682. November 2002.
50. M. Roth and R. Peikert. Flow Visualization for Turbomachinery Design. In *Proceedings IEEE Visualization '96*, pages 381–384, October 1996.
51. M. Roth and R. Peikert. A Higher-Order Method For Finding Vortex Core Lines. In *Proceedings IEEE Visualization '98*, pages 143–150, 1998.
52. Y. Rubner, C. Tomasi, and L. J. Guibas. A Metric for Distributions with Applications to Image Databases. 1998.
53. I. A. Sadarjoen and F. H. Post. Detection, Quantification, and Tracking of Vortices using Streamline Geometry. *Computers and Graphics*, 24(3):333–341, June 2000.
54. J. Sahner, T. Weinkauf, and H. C. Hege. Galilean Invariant Extraction and Iconic Representation of Vortex Core Lines. In *Proceedings of the Joint Eurographics - IEEE VGTC Symposium on Visualization (EuroVis 05)*, 2005.

55. G. Scheuermann, H. Hagen, H. Krüger, M. Menzel, and A. Rockwood. Visualization of Higher Order Singularities in Vector Fields. In *Proceedings IEEE Visualization '97*, pages 67–74, October 1997.
56. G. Scheuermann, H. Krüger, M. Menzel, and A. P. Rockwood. Visualizing Nonlinear Vector Field Topology. *IEEE Transactions on Visualization and Computer Graphics*, 4(2):109–116, April/June 1998.
57. G. Scheuermann and X. Tricoche. Topological Methods in Flow Visualization. In C.D. Hansen and C.R. Johnson, editors, *The Visualization Handbook*, pages 341–358. Elsevier, 2005.
58. S. Stegmaier and T. Ertl. A Graphics Hardware-based Vortex Detection and Visualization System. In *Proceedings IEEE Visualization 2004*, pages 195–202, 2004.
59. D. Sujudi and R. Haimes. Identification of Swirling Flow in 3D Vector Fields. Technical Report AIAA Paper 95–1715, American Institute of Aeronautics and Astronautics, 1995.
60. L. Sun, R Batra, X. Shi, and L. Hesselink. Topology Visualization of the Optical Power Flow through a Novel C-Shaped Nano-Aperture. In *Proceedings IEEE Visualization 2004*, pages 337–344, 2004.
61. A. Szymczak and J. Vanderhyde. Extraction of Topologically Simple Isosurfaces from Volume Datasets. In *Proceedings IEEE Visualization 2003*, pages 67–74, 2003.
62. C. K. Tang and G. G. Medioni. Extremal Feature Extraction From 3D Vector and Noisy Scalar Fields. In *Proceedings IEEE Visualization '98*, pages 95–102. IEEE, 1998.
63. A. Telea and J.J. van Wijk. Simplified Representation of Vector Fields. In *Proceedings IEEE Visualization '99*, pages 35–42, 1999.
64. H. Theisel. Designing 2D Vector Fields of Arbitrary Topology. *Computer Graphics Forum (Eurographics 2002)*, 21(3):595–595, 2002.
65. H. Theisel, C. Rössl, and H. Seidel. Using Feature Flow Fields for Topological Comparison of Vector Fields. In *Proceedings of the Conference on Vision, Modeling and Visualization 2003 (VMV-03)*, pages 521–528, November 19–21 2003.
66. H. Theisel, Ch. Rössl, and H.-P. Seidel. Combining Topological Simplification and Topology Preserving Compression for 2D Vector Fields. In *Pacific Graphics*, pages 419–423, 2003.
67. H. Theisel, Ch. Rössl, and H.-P. Seidel. Compression of 2D Vector Fields Under Guaranteed Topology Preservation. In *Eurographics (EG 03)*, volume 22(3) of *Computer Graphics forum*, pages 333–342, September 1–6 2003.
68. H. Theisel and H.-P. Seidel. Feature Flow Fields. In *Proceedings of the Joint Eurographics - IEEE TCVG Symposium on Visualization (VisSym 03)*, pages 141–148, 2003.
69. H. Theisel, J. Sahner, T. Weinkauf, H.-C. Hege, and H.-P. Seidel. Extraction of Parallel Vector Surfaces in 3D Time-Dependent Fields and Application to Vortex Core Line Tracking. In *Proceedings IEEE Visualization 2005*, pages 631–638, 2005.
70. H. Theisel, T. Weinkauf, H.-C. Hege, and H.-P. Seidel. Saddle Connectors–An Approach to Visualizing the Topological Skeleton of Complex 3D Vector Fields. In *Proceedings IEEE Visualization 2003*, pages 225–232, 2003.

71. H. Theisel, T. Weinkauf, H.-C. Hege, and H.-P. Seidel. Stream Line and Path Line Oriented Topology for 2D Time-Dependent Vector Fields. In *Proceedings IEEE Visualization 2004*, pages 321–328, 2004.
72. H. Theisel, T. Weinkauf, H.-C. Hege, and H.-P. Seidel. Topological Methods for 2D Time-Dependent Vector Fields Based On Stream Lines and Path Lines. *IEEE Transactions on Visualization and Computer Graphics*, 11(4), 2005.
73. H. Theisel, T. Weinkauf, H.-P. Seidel, and H. Seidel. Grid-Independent Detection of Closed Stream Lines in 2D Vector Fields. In *Proceedings of the Conference on Vision, Modeling and Visualization 2004 (VMV 04)*, pages 421–428, November 2004.
74. X. Tricoche, C. Garth, G. Kindlmann, E. Deines, G. Scheuermann, and H. Hagen. Visualization of Intricate Flow Structures for Vortex Breakdown Analysis. In *Proceedings IEEE Visualization 2004*, pages 187–194, 2004.
75. X. Tricoche, C. Garth, and G. Scheuermann. Fast and Robust Extraction of Separation Line Features. In *Proceedings of Seminar on Scientific Visualization 2003, Schloss Dagstuhl*, 2003.
76. X. Tricoche and G. Scheuermann. Continuous Topology Simplification of Planar Vector Fields. In *Proceedings IEEE Visualization 2001*, pages 159–166, 2001.
77. X. Tricoche, G. Scheuermann, and H. Hagen. A Topology Simplification Method For 2D Vector Fields. In *Proceedings IEEE Visualization 2000*, 2000.
78. X. Tricoche, G. Scheuermann, and H. Hagen. Scaling the Topology of Symmetric, Second-Order Planar Tensor Fields. In *Proceedings of NSF/DOE Lake Tahoe Workshop on Hierarchical Approximation and Geometrical Methods for Scientific Visualization*, 2001.
79. X. Tricoche, G. Scheuermann, and H. Hagen. Tensor Topology Tracking: A Visualization Method for Time-Dependent 2D Symmetric Tensor Fields. In *Computer Graphics Forum 20(3) (Eurographics 2001)*, pages 461–470, September 2001.
80. X. Tricoche, G. Scheuermann, and H. Hagen. Topology-Based Visualization of Time-Dependent 2D Vector Fields. In *Proceedings of the Joint Eurographics - IEEE TCVG Symposium on Visualization (VisSym-01)*, pages 117–126, May 28–30 2001.
81. X. Tricoche, G. Scheuermann, and H. Hagen. *Topology Simplification of Symmetric, Second-Order 2D Tensor Fields, Hierarchical and Geometrical Methods in Scientific Visualization.* 2003.
82. X. Tricoche, G. Scheuermann, H. Hagen, and S. Clauss. Vector and Tensor Field Topology Simplification on Irregular Grids. In *Proceedings of the Joint Eurographics - IEEE TCVG Symposium on Visualization (VisSym-01)*, pages 107–116. Springer-Verlag, May 28–30 2001.
83. X. Tricoche, T. Wischgoll, G. Scheuermann, and H. Hagen. Topology Tracking for the Visualization of Time-Dependent Two-Dimensional Flows. *Computers & Graphics*, 26(2):249–257, 2002.
84. T. Weinkauf, H. Theisel, H. C. Hege, and H.-P. Seidel. Boundary Switch Connectors for Topological Visualization of Complex 3D Vector Fields. In *Proceedings of the Joint Eurographics - IEEE TCVG Symposium on Visualization (VisSym 04)*, pages 183–192, 2004.
85. T. Weinkauf, H. Theisel, H.-C. Hege, and H.-P. Seidel. Topological Construction and Visualization of Higher Order 3D Vector Fields. *Computer Graphics Forum*, 23(3):469–478, 2004.

86. T. Weinkauf, H. Theisel, H. C. Hege, and H.-P. Seidel. Extracting Higher Order Critical Points and Topological Simplification of 3D Vector Fields. In *Proceedings IEEE Visualization 2005*, pages 559–566, 2005.
87. T. Wischgoll and G. Scheuermann. Detection and Visualization of Closed Streamlines in Planar Fields. *IEEE Transactions on Visualization and Computer Graphics*, 7(2), 2001.
88. T. Wischgoll and G. Scheuermann. Locating Closed Streamlines in 3D Vector Fields. In *Proceedings of the Joint Eurographics - IEEE TCVG Symposium on Visualization (VisSym 02)*, pages 227–280, May 2002.
89. T. Wischgoll, G. Scheuermann, and H. Hagen. Tracking Closed Streamlines in Time Dependent Planar Flows. In *Proceedings of the Vision Modeling and Visualization Conference 2001 (VMV 01)*, pages 447–454, November 21–23 2001.
90. X. Zheng and A. Pang. Topological Lines in 3D Tensor Fields. In *Proceedings IEEE Visualization '04*, pages 313–320, 2004.
91. X. Zheng and A. Pang. 2D Asymmetric Tensor Fields. In *Proceedings IEEE Visualization 2005*, pages 3–10, 2005.
92. X. Zheng, B. Parlett, and A. Pang. Topological Lines in 3D Tensor Fields and Discriminant Hessian Factorization. *IEEE Transactions on Visualization and Computer Graphics*, 11(4):395–407, July 2005.
93. X. Zheng, B. Parlett, and A. Pang. Topological Structures of 3D Tensor Fields. In *Proceedings IEEE Visualization 2005*, pages 551–558, 2005.

Topology-guided Visualization of Constrained Vector Fields

Ronald Peikert and Filip Sadlo

Computer Graphics Laboratory, Computer Science Department,
ETH Zurich, Switzerland
{peikert,sadlo}@inf.ethz.ch

Summary. In this study we explore ways of using precomputed vector field topology as a guide for interactive feature-based visualization of flow simulation data. Beyond streamline seeding based on critical points, we focus mainly on computing special stream surfaces related to critical points and periodic orbits. We address the special case of divergence-free vector fields which is often met in practical CFD data, and we extend the topological analysis to no-slip boundaries by treating 3D velocity and 2D wall shear stress in a unified way. Finally we apply the proposed techniques to flow simulation data and demonstrate their usefulness for the purpose of studying recirculation and separation phenomena.

1 Introduction

Vector field topology as a means to visualize the structure of fluid flow has been introduced by Helman and Hesselink [6]. A first generation of topology-based visualization methods locates, classifies, and displays critical points of the given vector field as point icons. Sophisticated icons can convey various information on the local topology and geometry of the flow [3]. Beyond critical points, periodic orbits can be located [19] and classified based on their Poincaré maps. Another use of Poincaré maps is to include them into 3D visualizations for a better understanding of the flow near the periodic orbit [8]. Finally, the topological skeleton of the vector field is obtained by computing all critical points and periodic orbits together with their stable and unstable manifolds, i.e. the union of streamlines converging in positive or negative time to the critical point or periodic orbit.

The striking property of these direct topological methods is that they are fully automatic and free of tuning parameters. A practical limitation is however that for many kinds of vector field data the topology is far too rich to be displayed in full detail. This led to concepts such as topological simplification [9, 16]. The stable or unstable 2D manifold of a 3D saddle point is a particularly interesting feature as it indicates a local flow separation. However, displaying a larger number of such stream surfaces leads to occlusion problems. Again, simplification is needed, and a possible solution is to only display their intersection curves, known as saddle connectors [17] or as heteroclinic and homoclinic orbits.

When considering the use of vector field topology for visualizing CFD data, it has to be kept in mind that topological features are not the final result an engineer or scientist wants to see. The topological analysis can, however, be a valuable first step to be followed by other visualization techniques. One possible strategy is to use topology for segmenting a vector field into regions of similar flow. This is particularly successful in 2D, while in 3D the notion of segmentation must be somehow relaxed to a more local property [10].

A second approach is to use topological features as guides for a different type of visualization. For example, a region-of-interest can be defined or a set of streamlines can be seeded [20] based on topological features. If an interactive, explorative type of visualization is pursued, visual clutter can usually be avoided, so that simplification is often not needed, even when stream surfaces are used for the visualization.

In this work we focus mainly on 2D manifolds of 3D saddles and saddle type periodic orbits. We believe that compared to arbitrarily chosen stream surfaces, such 2D manifolds can be more expressive and in most cases also of a simpler shape. In particular, recirculation zones and separation surfaces are well suited for this type of visualization. The underlying idea of visualizing topologically meaningful stream surfaces and their relationship to topological features has previously been used by Garth et al. [4] in their visualization of a vortex breakdown in the flow over a delta wing.

Computing stream surfaces in the vicinity or even converging to singularities, requires robust algorithms. The classical stream surface algorithm is that of Hultquist [7]. Here, the stream surface is generated by integrating a sequence of discretized streamlines and triangulating between them where appropriate. Triangle shape is optimized by choosing the shorter of the two possible edges in the process of triangulating between two streamlines. Triangle size is controlled by seeding new streamlines or stopping streamlines. This basic algorithm can be implemented with a depth-first strategy. However, to evaluate the criteria for adding or stopping a streamline, it is more convenient to used a breadth-first strategy where a current "front" is used. Garth et al. [4] added a refinement criterion based on the angle between adjacent segments of the front. Theisel et al. [17] remarked that Hultquist's algorithm fails if the tangents of the front are almost in the direction of the vector field, a situation which can arise e.g. near critical points or periodic orbits. They use as an initial front a line perpendicular to the vector field. This way, even tightly spiralling streamlines can be handled. However, the choice of the line is critical to avoid cracks or multiple coverings. Also, this approach produces spurious internal boundaries which have to be postprocessed for a correct result.

Vector field topology requires differentiable 2D or 3D vector fields. Usually, no further restriction is made for the vector fields. This is appropriate in the context of dynamical systems [2], which was the original application of vector field topology. Vector fields arising in physics, however, are often known to be divergence-free or irrotational or both. In Sec. 2 we will explore some of the implications of zero divergence to vector field topology and its application to the visualization of flow structures. In most CFD simulations, no-slip boundary conditions are imposed on some of the boundaries. Extending vector field topology to no-slip boundaries is the topic of Sec. 3. And finally, in Sec. 4 we will discuss some applications.

2 Topology of divergence-free vector fields

The case of a divergence-free (sometimes called solenoidal) vector field is particularly important in fluid dynamics. Examples of divergence-free vector fields are: velocity fields in hydrodynamics, vorticity fields, magnetic fields. Further divergence-free fields may be obtained by multiplying a given vector field (having neither sources nor sinks) with an appropriate scalar field. This is based on the fact that multiplication with a nonvanishing scalar field does not change the topology. As an example, the momentum field has the same topology as the velocity field, because they are identical up to a nonvanishing factor, the density. If the velocity field for instance is a steady solution of the compressible continuity equation $\frac{\partial \rho}{\partial t} + \nabla \cdot (\rho \mathbf{u}) = 0$, then the momentum field would be divergence-free.

The special case of divergence-free vector fields has an effect on the analysis of critical points. Asimov [1] mentions that in 2D and 3D divergence-free vector fields sources and sinks are not possible, but any types of saddles are. And in the 2D case, there is a new structurally stable type of critical points, namely the center. The center is said to have *constrained structural stability*. The center has the property that in a neighborhood, all streamlines are closed.

A similar analysis as for critical points can be done for periodic orbits (closed streamlines) in divergence-free 3D vector fields. Periodic orbits are of interest as they can indicate recirculation zones. Many properties of the periodic orbit can be studied in two dimensions by computing a Poincaré map. This is done by selecting a surface patch $\mathcal{S}$ which is everywhere transversal to the vector field. If sufficiently small, this so-called Poincaré section $\mathcal{S}$ is intersected by the periodic orbit in a single point. For a sufficiently close point $\mathbf{x} \in \mathcal{S}$, the Poincaré map $\mathcal{P}(\mathbf{x})$ is then defined as the first intersection of the streamline seeded at $\mathbf{x}$ with $\mathcal{S}$.

Periodic orbits are called *hyperbolic* if the eigenvalues of the linearization $\mathbf{P}$ of $\mathcal{P}$, the so-called Floquet multipliers, lie off the complex unit circle. According to Asimov [1], hyperbolic periodic orbits can be classified into sources, sinks, saddles, twisted saddles, spiral sources and spiral sinks depending on the Floquet multipliers.

The Poincaré map $\mathcal{P}$ has a fixed point where it is intersected by the periodic orbit. For the eigenvalue analysis, $\mathcal{P}$ is now linearized in a neighborhood of a fixed point. This linearized map $\mathbf{P}$ takes an infinitesimal circle centered at the fixed point to an ellipse with the same center. If the velocity field is divergence-free and thus volume preserving, the fluxes through the circle and the ellipse are equal. The flux is the integral of the normal velocity over the circle or ellipse. The normal velocity can be linearized as well, and because of symmetry, it can be replaced by its average. It follows that $\mathbf{P}$ must be area conserving, i.e. has a determinant of one. The sign is positive because a Poincaré map always conserves orientation.

2.1 Source and sink periodic orbits

It is now easy to see that periodic orbits of type source or sink are not possible for a divergence-free vector field. In the case of a source (either node source or spiral source), both eigenvalues lie outside of the complex unit circle. Hence, the determinant of $\mathbf{P}$ has absolute value greater than one, meaning that the area of an infinitesimal circle is not conserved under $\mathcal{P}$. The same can be concluded for sinks.

2.2 Saddle and twisted saddle periodic orbits

Periodic orbits of type saddle or twisted saddle are possible in divergence-free vector fields. Such periodic orbits are particularly suitable for visualization because they have a stable and an unstable manifold which are stream surfaces converging to the periodic orbit in positive or negative time. The nice property of these manifolds is that they "return to themselves" when following the periodic orbit for a full turn. This means, if a streamline is seeded on the intersection of the manifold with a Poincaré section and sufficiently close to the periodic orbit, it will return to the same intersection curve. If the seed curve is reduced to an infinitesimal line segment, its behavior is given by the eigenvalues of $\mathbf{P}$. If both eigenvalues are positive, the generated stream surface band returns untwisted to the Poincaré section. It may have done an integer number of full (360 degrees) so-called extrinsic twists. And it can shrink or stretch, depending on the eigenvalue associated to the eigenvector aligned with the seed line. If both eigenvalues are negative, the stream surface band does an additional half twist. In our case of divergence-free vector fields the product of the two eigenvalues equals one because of the above-mentioned conservation of area. Because of their property to return to the seed curve, (un-)stable manifolds are the ideal stream surfaces to depict the local behavior of the field near the periodic orbit.

2.3 Center periodic orbits

If a periodic orbit in a divergence-free vector field has complex eigenvalues of $\mathbf{P}$ its type can be neither spiral source nor spiral sink. It must be the in-between case with eigenvalues on the complex unit circle. This is not a hyperbolic case, but has the constrained structural stability similar to that of center critical points in 2D fields. By analogy, we call it a *center periodic orbit.*

The linearized Poincaré map $\mathbf{P}$ of such a periodic orbit has complex eigenvalues and a determinant of one. It can therefore be written as $\mathbf{P} = \mathbf{TRT}^{-1}$ where $\mathbf{R}$ is a pure rotation. It follows that $\mathbf{T}$ applied to an infinitesimal circle is an ellipse which is invariant under $\mathbf{P}$. This means that a stream surface seeded at this ellipse returns to the ellipse after following the periodic orbit for a full turn. The same idea can be used for finding finite invariant tori. The goal is here to find a closed seeding curve in the Poincaré section which is invariant under $\mathcal{P}$. As an initial guess a scaled version of the infinitesimal ellipse can be used. If starting from this an invariant seeding curve can be found, the problem is solved. However, we found that in practice this is a numerically challenging problem.

3 Topology near no-slip boundaries

3.1 Velocity and wall shear stress

By definition, a critical point is an isolated singularity of the vector field. Vector topology does not treat extended singularities. However, these occur in practical vector fields having solid boundaries with associated no slip boundary conditions. The velocity field $\mathbf{u}(\mathbf{x})$ itself is zero on such a boundary, but by using the unsigned distance to the boundary as a scalar field $s(\mathbf{x})$, it can be written as a product

$$\mathbf{u}(\mathbf{x}) = s(\mathbf{x})\tilde{\mathbf{u}}(\mathbf{x}), \tag{1}$$

where the vector field $\tilde{\mathbf{u}}(\mathbf{x})$ can be assumed to exist also on the boundary and to be nondegenerate there.

From the divergence-free criterion follows for points on the boundary:

$$0 = \nabla \cdot (s\tilde{\mathbf{u}}) = (\nabla s) \cdot \tilde{\mathbf{u}} + s\,(\nabla \cdot \tilde{\mathbf{u}}) = (\nabla s) \cdot \tilde{\mathbf{u}} \tag{2}$$

which means that on the boundary the field $\tilde{\mathbf{u}}$ has no normal component. In terms of vector field topology this means that no streamline of $\tilde{\mathbf{u}}$ ever passes from the solid boundary to the interior or vice versa.

If Eq. 2 holds, then on the boundary, $\tilde{\mathbf{u}}$ is related to the wall shear stress τ_w by $\tau_w = \mu\tilde{\mathbf{u}}$ where μ is the kinematic viscosity of the fluid. Because of this proportionality $\tilde{\mathbf{u}}$ has the same topology as the 2D field of wall shear stresses. At interior points, s is nonzero and therefore $\tilde{\mathbf{u}}$ has the same topology as $\mathbf{u}$. Hence, the field $\tilde{\mathbf{u}}$ nicely combines the wall shear field with the interior velocity field. However, this relies on the divergence-free property of the vector field. In the general case the field $\tilde{\mathbf{u}}$ has a normal component on the solid boundary. It can not be used to produce the topology of both the velocity field and the wall shear field. Of course the two vector fields could be blended, but then the topology of $\mathbf{u}$ may not be conserved.

Returning to the case of a divergence-free field, we saw that no streamline of $\tilde{\mathbf{u}}$ passes from the interior to the boundary. But there may be convergence towards critical points on the solid boundary, which are 3D saddle points having two of its eigendirections along the boundary surface. Also, convergence towards periodic orbits on the boundary is possible.

The advantage of using the field $\tilde{\mathbf{u}}$ is that it is no more necessary to extract both 2D and 3D critical points (with possible consistency issues). Critical points on the boundary are now regular 3D critical points. In the special case of $\mathbf{u}$ being divergence-free, sources and sinks can be excluded due to structural stability and the fact that $\tilde{\mathbf{u}}$ has the same topology as $\mathbf{u}$. Consequently, such critical points must be saddles or spiral saddles. Furthermore, by Eq. 2 their two-dimensional stable or unstable manifolds lie completely on the boundary. The eigenvalue belonging to the remaining eigenvector is real-valued. Its sign determines whether the point is on a separation line (positive sign) or a reattachment line (negative sign).

In discrete data, dividing by s has the drawback that due to interpolation inside the cells the topology is changed. A better strategy is to use the original field $\mathbf{u}$ for computing and analyzing the critical points in all cells which are not adjacent to no-slip boundaries. Only for computing the topology in the first layer of cells at the boundary, the modified field $\tilde{\mathbf{u}}$ is actually needed. The following steps are performed for cells adjacent to no-slip boundaries:

1. On interior nodes: compute $\tilde{\mathbf{u}}$ by dividing $\mathbf{u}$ by the wall distance.
2. On boundary nodes: interpolate $\mathbf{u}$ on two points on the boundary normal, compute $\tilde{\mathbf{u}}$, and linearly extrapolate to the boundary node.
3. Find critical points on the cell faces on the no-slip boundary. Use a 2D algorithm for finding the critical points, but classify them as 3D critical points.

3.2 Critical points on no-slip boundaries

Critical points on no-slip boundaries are important features for the study of flow separation. By applying the 3D classification, we will now concentrate on saddles and spiral saddles and ignore sinks and sources. These are of minor interest for the study of flow separation, and in divergence-free vector fields they do not occur.

Attracting (i.e. 2:1) saddles and spiral saddles have their stable manifold completely on the boundary. Any boundary curve of this manifold is a separation line. Similarly, repelling (1:2) saddles and spiral saddles have their unstable manifold on the boundary, so any boundary curve of it is a reattachment line. The case of spiralling separation (called tornado-type separation in [11]) has not been discussed much in the visualization community.

A pattern we encountered often consists of a pair of spiral saddles, one of them in the interior and one on a solid boundary (see Fig. 1, points C_1 and C_2, respectively). They are rotating in the same sense and mark a recirculation area.

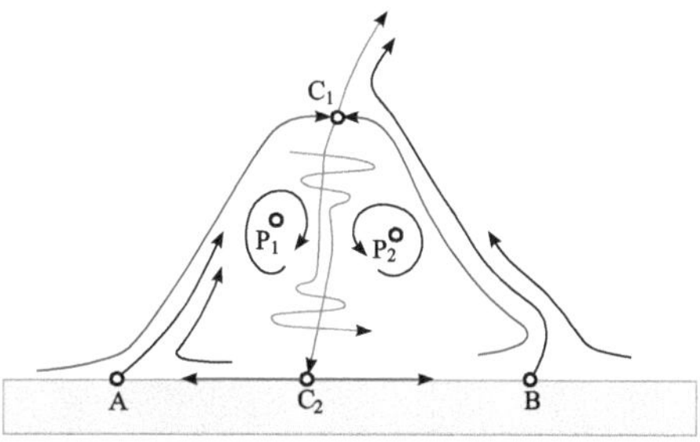

Fig. 1. Sketch of typical recirculation zone with two critical points of type spiral saddle (C_1, C_2) and one periodic orbit (P_1, P_2) involved. 1D manifolds (red curves) nearly meet. 2D manifolds (shown as blue curves) have a strong spiralling component.

The 1D manifolds nearly meet, while the 2D manifold of C_1 encloses the recirculation zone. This stream surface is not closed, so recirculation is not perfect. Within the recirculation zone there is a periodic orbit (P_1 and P_2). Finally, the points A and B appearing as saddles in the planar section, seem to indicate a separation line. However, these points are topologically nothing special, they are just the points where the skin friction line is intersected orthogonally by the planar section. This means that many nearby skin friction lines can be regarded as separation lines.

Surana et al. in their exact theory of flow separation [11] suggest as a criterion "strong hyperbolicity", i.e. large absolute eigenvalues of the saddles in the orthogonal section. An alternative and purely topological definition would be to pick the boundary curve of the unstable manifold of C_2 (which lies entirely on the solid boundary). In general this is composed of separatrices of nearby saddle points on the solid boundary.

4 Applications

4.1 Pelton turbine

A first application is a dataset resulting from a CFD simulation by VA Tech Hydro for the study of a Pelton turbine with the primary goal to optimize the stability of the water jets. The jets generated in the injectors (Fig. 2) must have a temporally stable circular cross section in order to optimally hit the runner buckets. Quality of the jets is mainly affected by vortices evolving in the outer ring where the water is deflected into the injectors. In Fig. 3 taken near the first of six injectors, the red stream surface shows the separation vortex arising because the flow does not follow the boundary. The yellow stream surface shows a smaller scale tornado-type separation.

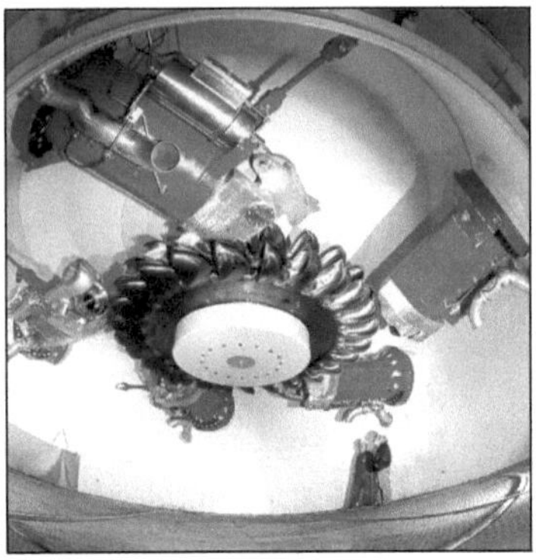

Fig. 2. Pelton turbine with five injectors (colorplate on p. 202).

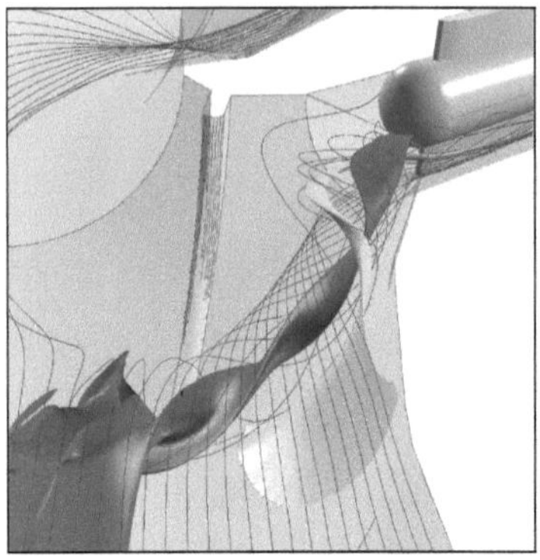

Fig. 3. Two vortices extending from the ring distributor into the first (of six) injectors (colorplate on p. 202).

Inspecting the nearby critical points reveals that there is a pair of spiral saddles in this region, one of them is on the no-slip boundary (upper right in Fig. 4). A quick exploration by integrating a streamline forward and backward from seed points near the critical points gives an idea of the stable and unstable manifolds of the two spiral saddles (Fig. 5).

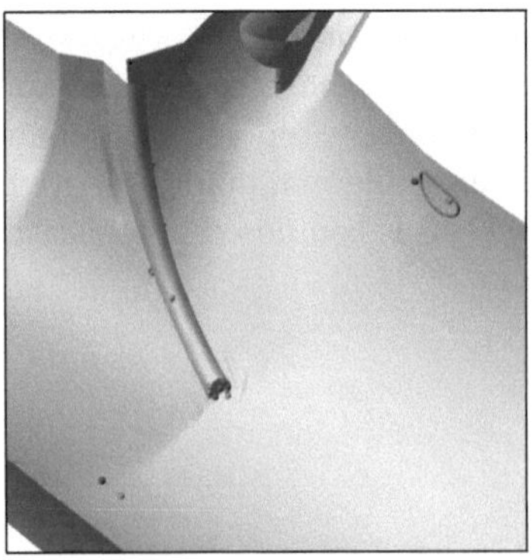

Fig. 4. Extracted interior (blue) and boundary (red) critical points. Periodic orbit (magenta). (Colorplate on p. 202.)

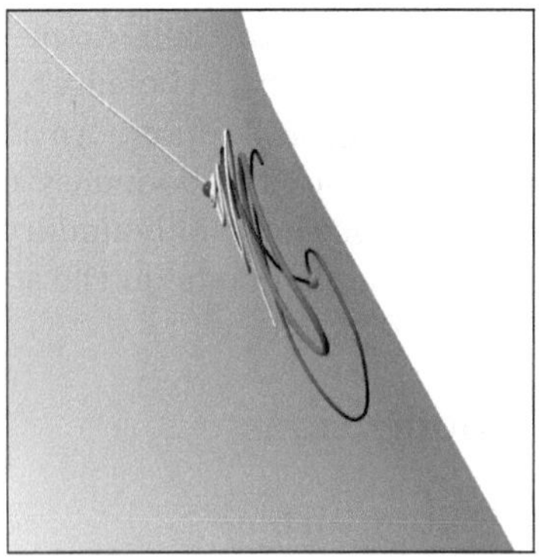

Fig. 5. Streamlines seeded near the boundary critical point (black) and the interior critical point (white). (Colorplate on p. 202.)

Consistent with the situation sketched in Fig. 1, the stable manifold of the interior critical point encloses the recirculation zone (Fig. 6). The recirculation zone contains a single periodic orbit which is of twisted saddle type (Fig. 7). In this case, the stable and unstable manifolds of the periodic orbit are classical Möbius strips with a half twist and no further extrinsic twisting.

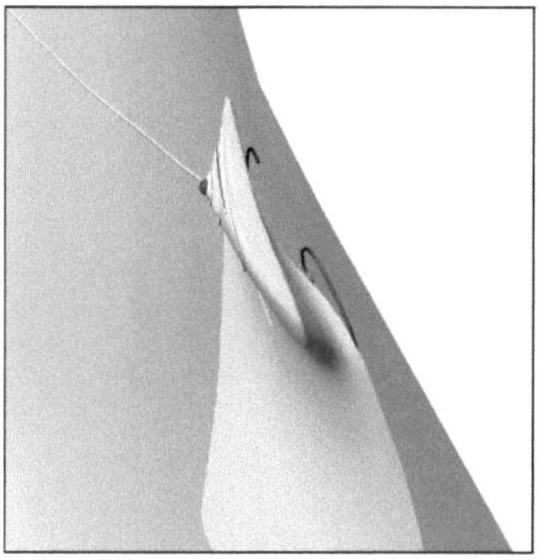

Fig. 6. Stable manifold of interior critical point (colorplate on p. 203).

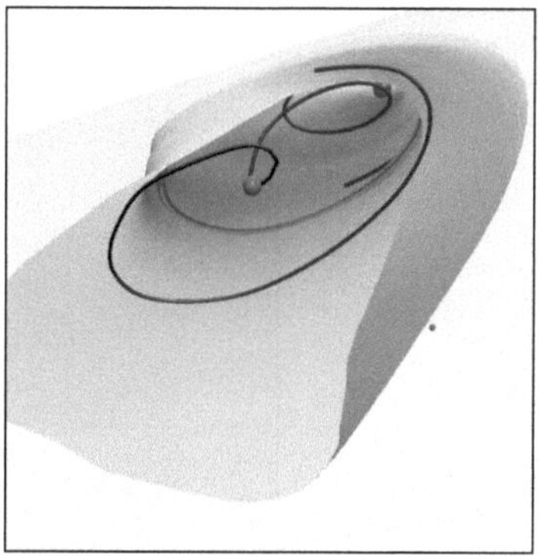

Fig. 7. View from the wall with stable and unstable manifolds of the periodic orbit (red and blue stream surfaces). (Colorplate on p. 203.)

Similar flow patterns as near the first injector also appear near the third and fifth injector. In all cases, a periodic orbit of type twisted saddle can be observed. However, in the case of the third injector, the eigenvalues are relatively close to -1, which suggests that instead of the twisted saddle, the center type (with a rotation angle close to 180 degrees) would be possible as well for slightly different data.

4.2 Draft tube

As a second application, we examined the flow in the incompressible CFD simulation of a Francis draft tube. The design of the draft tube is such that in its lower part it is split into two channels. As observed in the simulation data, the right channel exhibits significantly stronger vortices. For topological interests we picked one of the strong vortices extending horizontally and almost orthogonally to the primary flow direction. The transient simulation of this vortex consists of 3 interesting and quite steady phases: First there is a vortex breakdown bubble of the unstable 2D manifold of a spiral saddle, enclosed in the stable 2D manifold of a spiral saddle. Then the bubble collapses and the stable manifold develops into a vortex breakdown bubble. Finally this bubble collapses too, leaving a common vortex. We have chosen a time step of the first phase where the vortex breakdown bubble is quite steady and hence the examination of its instantaneous topology should reveal some of its properties. Additionally, as reported by [12], vortex breakdown bubbles are not necessarily the result of unsteady flow behavior.

Fig. 8. Tornado-type separation and vortex in the draft tube dataset. Stream surface (transparent blue) starts at saddle and goes upstream enclosing a vortex breakdown bubble (blue streamline) containing a periodic orbit (red). Critical points (red) and vortex core lines (green). (Colorplate on p. 203.)

Fig. 8 gives a view from top on the flow going from left to right. It shows (from bottom to top) a tornado-type separation with a critical point on the (outer) wall, and a vortex core line that connects to that critical point and extends across the channel into the part where the two channels merge. There is a recirculation region identified as a vortex breakdown bubble with a critical point at its bottom and a periodic orbit inside it. Another critical point resides above the bubble where the detected core line is disrupted. The stable 2D manifold of that saddle is visualized by an upstream surface that encloses the vortex breakdown bubble and approaches the wall.

Unperturbed vortex breakdown bubbles (Fig. 9) are axisymmetric and consist of nested invariant tori. In the real world [18, 14], they contain regions of chaotic dynamics with possible islands of stability and KAM tori (impermeable) or Cantori (permeable) separating the regions (Fig. 10). We refer the reader to the paper of Sotiropoulos et al. [13] for details. In the field of visualization, vortex breakdown bubbles have been studied recently ([15, 5, 4]).

In our case, a vortex breakdown bubble containing a periodic orbit has been identified (Fig. 10). It seems that the stable 2D manifold of the upper saddle (transparent stream surface in Fig. 8) marks the end of the recirculation

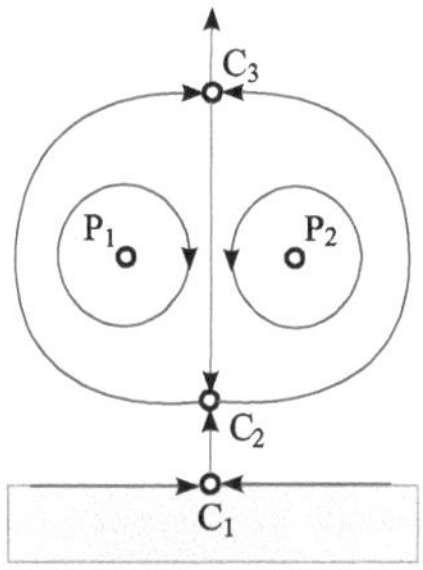

Fig. 9. Sketch of ideal (unperturbed) vortex breakdown bubble. Three critical points (C_1, C_2, C_3) and one periodic orbit (P_1, P_2).

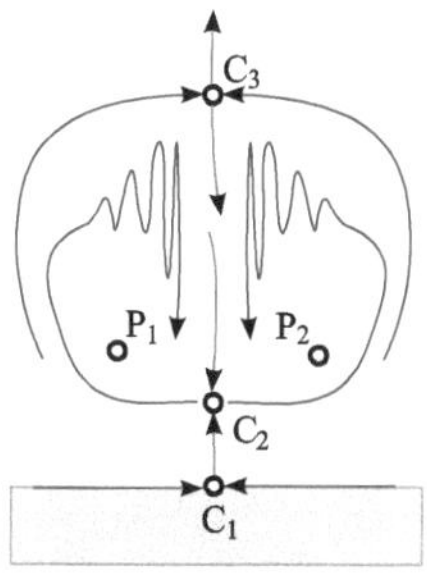

Fig. 10. Sketch of real-world (perturbed) vortex breakdown bubble. No intersection of the two 2D manifolds due to nonzero divergence in this region.

region since Spohn et al. [12] report that vortex breakdown bubbles exhibit permanent inflow and outflow at the downstream tail. The fact that the two manifolds do not intersect is due to the inaccuracy of the simulation leading to nonzero divergence in this region.

Computing the vortex breakdown bubble as a stream surface seems impossible with Hultquist-type algorithms due to the complex folding and also due to the quasi-periodicity of the streamlines. Since a single streamline covers the toroidal stream surface densely, it can be seeded near the critical point and sampled on a voxel grid. The resulting field can then be visualized by an isosurface. To reduce aliasing effects and enhance resolution, a voxel value is not set in a binary manner when the streamline passes but computed based on coverage. An initial sequence of integration steps was not sampled in order to avoid an isolated spiral from the saddle point to the unstable manifold of the bubble.

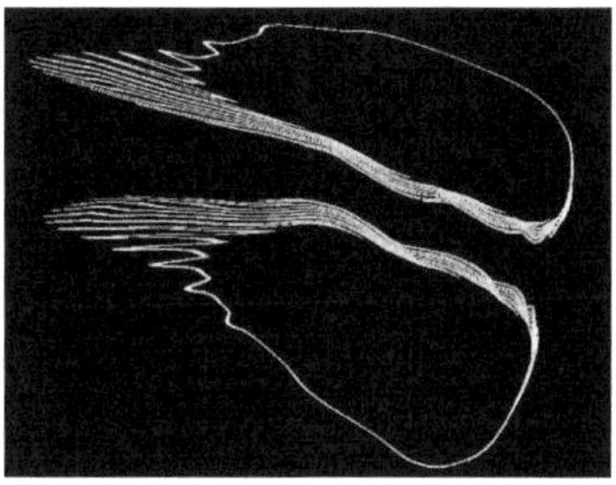

Fig. 11. Slice of the voxel field that sampled a single streamline.

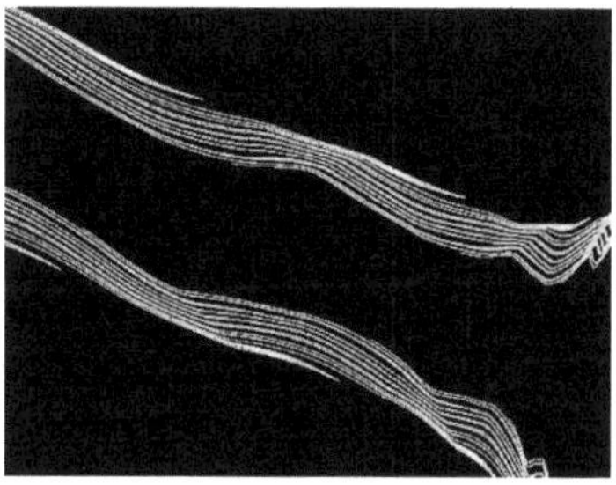

Fig. 12. Detail sampled at higher voxel resolution.

Fig. 11 shows a slice of the resulting voxel field after tracing the particle for 10^9 time steps. Its resolution is $750 \times 600 \times 600$ and it spans the complete bubble. Fig. 12 shows a finer sampling of a subregion. This makes the massive folding of the surface visible. Fig. 13 and 14 show an isosurface of the voxel field with the complete bubble. The isolevel was chosen to be 5% instead of 50% in order to avoid unmanageably many triangles in the fine folds. By adding a Gaussian smoothing step, we were able to cut down the the triangle count to about 20 million.

Fig. 13. Isosurface of the voxel-sampled single streamline on the surface of the breakdown bubble.

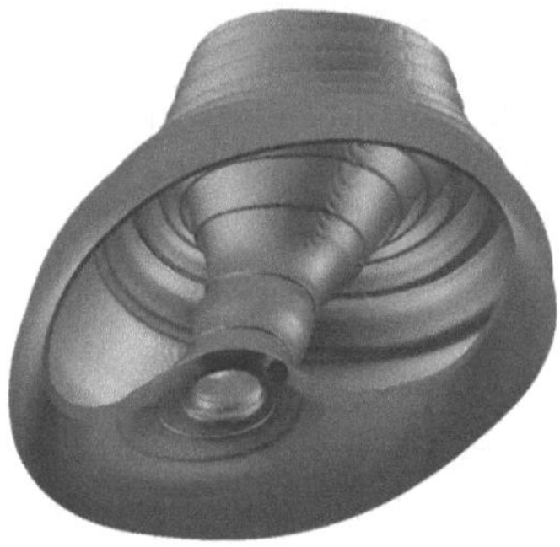

Fig. 14. Same isosurface clipped for view to the inside.

5 Conclusion

We gave examples of flow features in real CFD datasets which can be nicely illustrated by 2D manifolds of 3D saddles. Our experience showed that stream surface integration gets particularly challenging for these special cases of stream surfaces. Very robust stream surface algorithms are required which can cope with situations such as tightly winding spirals occurring in 2D manifolds of spiral saddles or saddle type periodic orbits. This issue is worth being addressed in further work, and as the ultimate goal in this line of research we see a stream surface algorithm which is fully "topology aware", i.e. which behaves correctly when integration approaches any kind of topological feature. The authors acknowledge VA Tech Hydro for the simulation data. This work was funded by Swiss Commission for Technology and Innovation grant 7338.2 ESPP-ES.

References

1. Asimov, D.: Notes on the Topology of Vector Fields and Flows. Tech. Report RNR-93-003, NASA Ames Research Center (1993)

2. Guckenheimer, J., Holmes, P.: Nonlinear Oscillations, Dynamical Systems and Bifurcations of Vector Fields. Springer, New York (1983)
3. Globus, A., Levit, C., Lasinski, T.: A tool for visualizing the topology of three-dimensional vector fields, In Proc. IEEE Visualization 91, 33-40 (1991)
4. Garth, C., Tricoche, X., Salzbrunn, T., Bobach, T., Scheuermann, G.: Surface Techniques for Vortex Visualization. In: Proceedings of VisSym 2004, 155–164 (2004)
5. Garth, C., Tricoche, X., Scheuermann, G.: Tracking of Vector Field Singularities in Unstructured 3D Time-Dependent Datasets. In: Proc. IEEE Visualization 2004, 329–336 (2004)
6. Helman, J.L., Hesselink, L.: Representation and Display of Vector Field Topology in Fluid Flow Data Sets, Computer, **22(8)**, 27–36 (1989).
7. Hultquist, J.P.M.: Constructing stream surfaces in steady 3D vector fields, In: Proceedings of the 3rd conference on Visualization '92, 171–178 (1992)
8. Löffelmann, H., Kucera, T., Gröller, E.: Visualizing Poincaré Maps Together With the Underlying Flow. In: Hege, H.C., Polthier, K. (eds.), Proceedings of the International Workshop on Visualization and Mathematics '97 315–328 (1998)
9. de Leeuw, W., van Liere, R.: Collapsing flow topology using area metrics, In: Proc. IEEE Visualization 99, 149-354 (1999)
10. Mahrous, K. Bennett, J., Scheuermann, G., Hamann, B. Joy, K.I.: Topological Segmentation in Three-Dimensional Vector Fields, IEEE Transactions on Visualization and Computer Graphics, **10(2)**, 198–205 (2004)
11. Surana, A., Grunberg, O., Haller, G.: Exact theory of threedimensional flow separation. Part I. Steady separation, J. Fluid Mech., **564**, 57–103 (2006)
12. Spohn, A., Mory, M, Hopfinger, E.: Experiments on vortex breakdown in a confined flow generated by a rotating disk, J. Fluid Mech., **370**, 73-99 (1998)
13. Sotiropoulos, F., Ventikos, Y., Lackey, T.: Chaotic advection in three-dimensional stationary vortex-breakdown bubbles: Sil'nikov's chaos and the devil's staircase, J. Fluid Mech., **444**, 257-297 (2001)
14. Thompson, M. Hourigan, K.: The sensitivity of steady vortex breakdown bubbles in confined cylinder flows to rotating lid misalignment, J. Fluid Mech., **496**, 129-138 (2003)
15. Tricoche, X., Garth, C., Kindlmann, G.L., Deines, E., Scheuermann, G., Ruetten, M., Hansen, C.D.: Visualization of Intricate Flow Structures for Vortex Breakdown Analysis. In: Proceedings of IEEE Visualization '04, 187–194, 2004.
16. Tricoche, X., Scheuermann, G., Hagen, H.: A topology simplification method for 2D vector fields, In: Proc. IEEE Visualization 2000, 359-366 (2000)
17. Theisel, H., Weinkauf, T., Hege, H.C., Seidel, H.P., Saddle Connectors - An Approach to Visualizing the Topological Skeleton of Complex 3D Vector Fields. In: Proc. IEEE Visualization 2003, 225–232 (2003)
18. Ventikos, Y.: The effect of imperfections on the emergence of three-dimensionality in stationary vortex breakdown bubbles, Physics of Fluids, L13-L16 (2002)
19. Wischgoll, T., Scheuermann, G.: Locating closed streamlines in 3D vector fields, In: D. Ebert, P. Brunet, I. Navazo (eds.), Data Visualisation 2002, Eurographics Association, 227–232 (2002)
20. Ye, X., Kao, D., Pang, A., Strategy for Scalable Seeding of 3D Streamlines, In: Proceedings of IEEE Visualization '05, Minneapolis (2005)

Scale-Space Tracking of Critical Points in 3D Vector Fields

Thomas Klein and Thomas Ertl

Institute for Visualization and Interactive Systems (VIS), Universität Stuttgart
{klein,ertl}@vis.uni-stuttgart.de

Summary. Scale-space techniques are very popular in image processing since they allow for the integrated analysis of image structure. The multi-scale approach enables one to distinguish between important features such as edges and small-scale features such as numerical artifacts or noise. In general, the same properties hold for vector fields such as flow data. Many flow features, e.g. vortices, can be observed on multiple scales of the data and also many features that can be detected are essentially artifacts of the employed interpolation scheme or originate from noise in the data. In this paper, we investigate an approach based on scale-space hierarchies of three-dimensional vector fields. Our main interest concerns how vector field singularities can be tracked over multiple spatial scales in order to assess the importance of a critical point to the overall behavior of the underlying flow field.

1 Introduction

The extraction of topological features has become a valuable tool for the analysis of vector field data that arise for example in computational or experimental fluid dynamics research. Vector field singularities in combination with a set of feature lines or surfaces, such as separatrices, provide the means of describing the qualitative behavior of a flow field in a strongly condensed form via the so-called topological skeleton. This enables one to cope with the ever growing dataset sizes in today's CFD applications.

Since typical datasets originating from simulations or real world measurements contain structures of different sizes or scales and different sets of features can be observed on certain ranges of scale, the notion of scale is an important concept in the analysis of the data. Therefore, an automatic or semi-automatic analysis tool for flow topology has to accommodate the inherent multi-scale nature of the underlying data.

Furthermore, many flow features, e.g. vortices, can be observed on multiple scales or resolutions of the data while many features that can be only detected at fine scales are essentially artifacts of the employed interpolation scheme or originate from noise in the data. Thus, a scale-aware technique can help to identify the overall structure of a given flow field and allows to distinguish the global structure from local effects, such as noise, and discretization or interpolation artifacts.

As in general the scales of the features are a priori not known, it seems to be reasonable to represent the data at multiple scales, successively eliminating fine-scale flow field structures. Nevertheless, in certain situations the scientist analyzing the data may be in possession of information that allows him to limit the parameter space to certain resolutions.

The concept of scale-space provides a well established framework to cope with these problems in a well defined way. Scale-space techniques have become very popular in image processing since they allow for the integrated analysis of the image structure. The multi-scale approach enables one to distinguish between important features, such as edges, and small-scale features, such as numerical artifacts or noise. By reason of this, scale-space techniques have been successfully applied to the automatic analysis of images in computer vision.

In this work we present our first attempt of a multi-scale topological analysis of flow data. We investigate an approach based on Gaussian scale-space hierarchies of three-dimensional vector fields. Our main interest concerns the tracking of important topological features, i.e. critical points, over multiple spatial scales in order to distinguish between local and global structures and behavior of the underlying vector field and numerical or noise-induced artifacts. This procedure is based on the assumption that fine scale structure and noise will be gradually eliminated on coarser scales while the dominating large scale flow structures will be persistent over multiple or all scales of the dataset. Thus, the scale length of the path or the number of scales over that a critical point can be tracked corresponds to the importance of this singularity to the overall topology of the flow.

2 Related Work

There is a huge body of literature dealing with the extraction and visualization of vector field topology. Since their first introduction to the context of visualization of two-dimensional vector field data by Helman and Hesselink [7, 8], topology-based methods have been established as one of the basic tools for flow analysis, and were soon generalized to three-dimensional fields, as well [8, 5].

Since then, many improvements and extensions have been published. The application of results from geometric calculus and Clifford algebra regarding vector field indices made it possible to detect and visualize also higher-order critical points and non-linear vector field topology. This has been investigated in detail by Scheuermann et al. [18, 19, 20] for two-dimensional flows and by Mann and Rockwood [15] for three-dimensional vector fields.

As visualizations of the topology of three-dimensional flow fields typically involve a number of separating stream surfaces they suffer from occlusion problems and accordingly tend to get visually cluttered. To deal with that problem, Theisel et al. introduced the concept of saddle connectors [22] and boundary switch connectors [26] as a method for the simplified visualization of the topological skeleton of complex three-dimensional vector fields. However, the visual complexity, even of these simplified visualizations, depends heavily on the number of critical points involved.

To deal with large numbers of critical points and the resulting complexity of topological skeletons of vector fields the multi-scale nature of the data has been addressed in different ways. The simplest approach is to use a low pass filtering technique to suppress small-scale features and noise in the data and to do the extraction of the critical points on the filtered data. However, this approach cannot guarantee the invariance of the position and classification of critical points in the dataset. Therefore, a number of topological simplification methods have been proposed, e.g. the work by de Leeuw van Liere [2] or Tricoche et al. [23]. In contrast to the approach we will describe in the following, however, these methods often depend on the availability of the complete topological skeleton, not merely the critical points.

On the other hand, scale-aware feature extraction techniques for scalar valued data have a long tradition in image processing and computer vision. It was in the context of pattern recognition when the concept of scale of features and the scale-space paradigm emerged first [9]. Starting with the work of Witkin [28] and Koenderink [10] the concept of Gaussian scale-space representations has gained much attention in the image processing literature. The analysis of the so-called *deep structure* of images by means of an investigation of a multi-scale image representation, has become a valuable tool for feature detection and extraction in scalar images [14]. Of special interest in our case is the work by Lindeberg [13] and by Florack and Kuijper [3, 12] on the scale-space behavior of critical points in scalar fields, since the approach we will describe in Sec. 5 is closely related to their work on critical curves, i.e. the trajectories of singularities in the gradient field of a scale-space image. Although, we will disregard their extensive work on bifurcations and degenerate singularities based on the framework of catastrophe theory [11].

The tracking of features in vector fields has also attracted the interest of a number of researchers in flow visualization, especially in the case of the analysis of time-dependent vector fields. A general overview of the state of the art in feature extraction and tracking in flow fields can be found in Post et al. [17]. Particularly concerned with the tracking of vector field singularities as a topological feature in time-dependent two- and three-dimensional datasets are the works by Tricoche et. al [24, 25], Theisel and Seidel [21], and Garth et al. [4]. Feature tracking in scale-space representations of vector fields has been investigated by Bauer and Peikert [1] for tracking the evolution of vortices in scale or time.

For the approach discussed in this work the papers by Theisel and Seidel [21] and Bauer and Peikert [1] are the two most relevant. In the first, the concept of feature flow fields is introduced. A vector field derived from the original flow field such that the evolution of the considered features is described by streamlines in this field. Thus, the feature tracking problem is reduced to simple streamline integration. We will come back to this technique in Sec. 5. In the second, vortex core lines based on the parallel vector operator are investigated. The construction of a linear scale-space representation of a flow field given on a three-dimensional unstructured grid is discussed and an algorithm for the tracking of the vortex core lines in scale-space based on a 4D surface extraction technique is presented. To our knowledge, this is the only work on scale-space feature tracking for flow visualization.

3 Scale-Space and Topology

3.1 Gaussian Scale-Space

As already mentioned, scale-space techniques have become very popular in image processing since they allow for the integrated analysis of image structure at all scales. Here we will briefly outline the basic idea in case of a three-dimensional vector field.

In the following we will assume $\mathbf{v} : M \to \mathbb{R}^3$ to be a vector field given on a three-dimensional domain $M \subseteq \mathbb{R}^3$. Introducing the scale parameter $\tau > 0$, the scale-space representation $\boldsymbol{\nu} : M \times \mathbb{R}^+ \to \mathbb{R}^3$ of $\mathbf{v}$ with $\lim_{\tau \to 0} \boldsymbol{\nu}(\mathbf{x}, \tau) = \mathbf{v}(\mathbf{x})$ defines an embedding of the original vector field in a one-parameter family of derived vector fields, resulting in a hierarchy consisting of subsequently simplified or smoothed versions of the original data. Thus, τ defines the scale-axis of the four-dimensional scale-space.

One possibility, and in fact the most often used, to generate such a one parameter family of vector field representations is the linear or Gaussian scale-space. Here the smoothing of the data is accomplished by the convolution $\boldsymbol{\nu}(\mathbf{x}, \tau) = \mathbf{v}(\mathbf{x}) * g(\mathbf{x}, \tau)$ of the original signal with Gaussian filter kernels

$$g(\mathbf{x}, \tau) = \frac{1}{(2\pi\tau)^{3/2}} \mathrm{e}^{-\frac{\|\mathbf{x}\|^2}{2\tau}} \qquad (1)$$

of increasing standard deviation $\sigma = \sqrt{\tau}$. Which is equivalent to the solution of the linear diffusion or heat equation

$$\partial_\tau \boldsymbol{\nu}(\mathbf{x}, \tau) = \frac{1}{2} \Delta \boldsymbol{\nu}(\mathbf{x}, \tau) \;, \qquad (2)$$

with initial condition

$$\boldsymbol{\nu}(\mathbf{x}, 0) = \mathbf{v}(\mathbf{x}) \;.$$

Where $\Delta = \frac{\partial^2}{\partial x^2} + \frac{\partial^2}{\partial y^2} + \frac{\partial^2}{\partial z^2}$ denotes the spatial Laplacian operator.

Hence, for a vector field given by data values on a discrete grid a Gaussian scale-space representation can be computed in at least two ways, by repeatedly filtering the data with sampled Gaussians of increasing variance and accordingly width or by solving the diffusion equation (2). Although, the second involves the solution of a partial differential equation, for large τ this methods becomes increasingly efficient compared to using sampled Gaussians.

For the computation of the scale-space representation of our vector fields we have implemented both methods, the repeated convolution with separated Gaussians of increasing variance and accordingly increasing support and the iterative solution of the linear diffusion equation using a solver based on the SOR method. In both cases we work on three-dimensional Cartesian grids although the second approach could be very easily extended to unstructured tetrahedral meshes.

3.2 Scale-Space Topology

The additional degree of freedom introduced by the scale parameter, in fact, provides further topological features. There exist transitions between the different scale levels that cause topological changes. Subtle changes of the scale parameter can cause structural changes in the system's topology. These so-called bifurcations or catastrophes can occur in every dynamic system that depends on a set of varying control parameters, such as scale in our case. Thereby, the local topology changes from one stable state to another via a transient unstable state.

These phenomena have been studied extensively in the theory of dynamic systems and can be described in the framework of catastrophe theory that deals with how critical points of a parameter dependent dynamic system will evolve when the control parameters are continuously changed. We will not go into detail here, but refer to the literature [6]. An extensive account of the behavior of critical points of scalar fields, i.e. singularities of their gradient fields, under Gaussian blurring, i.e. in linear scale-space, can be found, e.g., in the work by Florack and Kuijper [3, 12].

Many different types of possible bifurcations exist in a general dynamic system [6], but in flow topology one is mostly concerned with three general types of bifurcation events that can occur in the topology of parameter dependent vector fields: annihilations and creations of pairs of critical points and critical points changing their classification. The last, the so called Hopf bifurcation, describes for example the transition from a sink to a source or vice versa.

Unfortunately, the possibility of creation events in a linear scale-space representation seems to be contradictory to the goal of topology simplification. New (unstable) degenerate critical points can be created during the diffusion process that subsequently will lead to the creation of a new pair of critical points. This is known as a static fold bifurcation. However, since we are only interested in the behavior of the critical points that exist in the original dataset, i.e. on the finest scale, we will in the following neglect this type of bifurcation event.

4 Detection of Vector Field Singularities

When talking about vector field singularities or critical points we will always assume spatial isolated zeros of the vector field with non-singular Jacobian matrix, i.e. first order critical points. This poses in general no problem, as line or surface singularities always correspond to degenerated critical points, i.e. points where the Jacobian of the field does not have full rank. Obviously, this also excludes higher-order critical points that may be present in non-linear vector fields [18].

The detection of vector field singularities is in general a numerical ill-posed and challenging problem. Noise and numerical inaccuracies will lead to false positives and actual singularities might be missed due to deficiencies of the employed interpolation scheme. As for all topology-based methods finding a complete seed set of critical points is a crucial part of the algorithm.

We have implemented and compared two approaches for the detection of critical points on a regular three-dimensional grid, namely the linearization of the vector field using a tetrahedral decomposition of the grid and a method based on the computation of winding numbers using geometric calculus. Both methods have certain strengths and weaknesses.

Decomposing the grid cells into tetrahedra corresponds to effectively linearizing the vector field, thus higher-order critical points – points with index not equal to 0 or ± 1 – cannot be detected directly by this method. Such points either are missed or show up as pairs of neighboring first-order critical points [18]. Although, for each grid cell multiple tetrahedra have to be processed, the actual computation per tetrahedron boils down to solving a linear equation system for the barycentric coordinates of the critical point's position inside the cell. Therefore, this method is still the simplest and fasted way to compute vector field singularities as long as higher-order critical points can be neglected.

Second, we investigated the method described by Mann and Rockwood [15] that is based on computing the index of a critical point using winding numbers. At the moment this is regarded as the most general method, since it is not restricted to detecting first order critical points but can theoretically detect critical points of arbitrary index. Using results from geometric calculus Mann and Rockwood determine the index of a critical point by computing an integral over a closed surface. The result of this computation is always an integer value, that is either zero if no critical point is enclosed by the surface or equals the index of the critical point enclosed by the surface. Unfortunately, there are also some problems with this approach. First of all, the value of the integral in general may not reflect the actual situation. Since the indices of multiple singularities enclosed by the surface add up in the result of the integration, an index of zero does not necessarily signify that no point is enclosed. The same applies for values not equal to zero. The problem is how to choose the appropriate spatial resolution for the enclosing manifold. At first sight one cell or the surface defined by its six faces seems to be a good choice, since this defines the maximal resolution of our data. But, as can be easily seen, a single cell is not sufficient to identify, for example, a dipole, at least if bilinear interpolation is used for sampling the cell faces. Thus, expensive higher-order interpolation schemes have to be used. But even when using tricubic interpolation, we experienced singularities that are missed. Furthermore, in order to achieve a reasonable accuracy for the integration, i.e. indices that are close to integer values, the sampling grid on the cell faces has to be sufficiently fine, which makes the computation even more expensive. Therefore, this method is not feasible for real datasets.

5 Tracking of Critical Points in Scale Space

In this section we will be concerned with the development of an algorithm for tracking non-degenerate singularities in scale-space. A straight forward solution would be to detect all critical points on all scales and then try to connect them according to some simple correspondence criteria, such as spatial distance or classification. There are a number of problems with this approach. In many situations such simple criteria are not sufficient to decide whether a critical point detected on a coarser scale really corresponds to a spatially close point on the next finer scale. Furthermore, tracking based on feature correspondence depends somehow or other on the definition of certain threshold values, e.g. for the maximum distance of two features considered to be equal. Often it is not possible to specify such values for an unknown dataset in advance. Thus, a fully automatic analysis of the data is not possible. In the following we describe two approaches that do not require such thresholds. The first is based on the concept of feature flow fields. The second exploits the explicit knowledge of the scale-space trajectory of a critical point provided by the implicit function theorem. Both approaches reduce the feature tracking problem to a stream line integration problem in a derived vector field, as will be described in the following.

5.1 Critical Point Tracking Based on Feature Flow Fields

As introduced by Theisel and Seidel [21], a so-called feature flow field of a time-dependent three-dimensional vector field $\mathbf{u}(\mathbf{x}, t) = (u_1(\mathbf{x}, t), u_2(\mathbf{x}, t), u_3(\mathbf{x}, t))^T$ is a four-dimensional vector field $\mathbf{f}(\mathbf{x}, t) = (f_1(\mathbf{x}, t), \ldots, f_4(\mathbf{x}, t))^T$ derived from $\mathbf{u}(\mathbf{x}, t)$ such that the time evolution of the considered features is described by streamlines in $\mathbf{f}(\mathbf{x}, t)$. In the case of a critical point c this means each point on the streamline in $\mathbf{f}(\mathbf{x}, t)$ starting from c is also a critical point, i.e. the value of $\mathbf{u}(\mathbf{x}, t)$ must not change when traversing the streamline. Assuming a first order approximation of $\mathbf{u}(\mathbf{x}, t)$ around c this implies that $\mathbf{f}(\mathbf{x}, t) \perp \nabla u_i(\mathbf{x}, t)$ for $i = 1, 2, 3$. When we apply this to our scale-space representation $\boldsymbol{\nu}(\mathbf{x}, t)$ by identifying time t with scale τ, the scale-space feature flow field $\mathbf{F}(\mathbf{x}, \tau)$ is given by

$$\mathbf{F}(\mathbf{x},\tau) = \begin{pmatrix} -\det(\boldsymbol{\nu}_y(\mathbf{x},\tau),\boldsymbol{\nu}_z(\mathbf{x},\tau),\boldsymbol{\nu}_\tau(\mathbf{x},\tau)) \\ \det(\boldsymbol{\nu}_z(\mathbf{x},\tau),\boldsymbol{\nu}_\tau(\mathbf{x},\tau),\boldsymbol{\nu}_x(\mathbf{x},\tau)) \\ -\det(\boldsymbol{\nu}_\tau(\mathbf{x},\tau),\boldsymbol{\nu}_x(\mathbf{x},\tau),\boldsymbol{\nu}_y(\mathbf{x},\tau)) \\ \det(\boldsymbol{\nu}_x(\mathbf{x},\tau),\boldsymbol{\nu}_y(\mathbf{x},\tau),\boldsymbol{\nu}_z(\mathbf{x},\tau)) \end{pmatrix}, \tag{3}$$

where $\boldsymbol{\nu}_x(\mathbf{x},\tau))$, $\boldsymbol{\nu}_y(\mathbf{x},\tau)$, and $\boldsymbol{\nu}_z(\mathbf{x},\tau))$ are given by the columns of the spatial Jacobian $\boldsymbol{\nu}_\mathbf{x}(\mathbf{x},\tau)$ of the scale-space representation.

Thus, the feature tracking problem is reduced to a four-dimensional streamline integration problem starting at an initial set of seed points given by the critical points encountered on the finest scale.

5.2 Singularity Tracking Based on the Implicit Function Theorem

In this section, we will describe an alternative approach for critical point tracking that has been used, for example, by Lindeberg [13] for analyzing the behavior of local extrema in images under Gaussian blurring. Similar to the feature flow field approach, the basic idea here is also that using explicit knowledge about the actual trajectories of the critical points in scale-space can significantly improve the tracking results.

As we will show in the following, the evolution of a non-degenerate critical point in scale-space can be analyzed by means of the general implicit function theorem. Computing the trajectory of a critical point in scale-space can be regarded to be equivalent to finding the level set $H : \boldsymbol{\nu}(\mathbf{x},\tau) = \mathbf{0}$. Then, for a given scale-space critical point $(\mathbf{x}_0,\tau_0) \in H$ with non-singular matrix $\boldsymbol{\nu}_\mathbf{x}(\mathbf{x}_0,\tau_0)$, i.e. $\det(\boldsymbol{\nu}_\mathbf{x}(\mathbf{x}_0,\tau_0)) \neq 0$, the implicit function theorem states the following: In a local neighborhood of the critical point, $\boldsymbol{\nu}(\mathbf{x},\tau) = \mathbf{0}$ can be solved for $\mathbf{x}$. In other words, there exists a function $\mathbf{h}(\tau)$ with $\mathbf{x}_0 = \mathbf{h}(\tau_0)$ and $\boldsymbol{\nu}(\mathbf{h}(\tau),\tau) = \mathbf{0}$. Hence, the path of a critical point in the four-dimensional scale-space is a one-dimensional manifold, i.e. a curve. Although, it is not guaranteed that there exists an explicit representation of $\mathbf{h}$, the tangent of the curve in $(\mathbf{x}_0,\tau_0)$ can be always computed as

$$\mathbf{h}'(\tau_0) = -(\boldsymbol{\nu}_\mathbf{x}(\mathbf{x}_0,\tau_0))^{-1}\boldsymbol{\nu}_\tau(\mathbf{x}_0,\tau_0)\,. \tag{4}$$

Repeating this argument, the integration of the path of the singularity through scale-space can then be accomplished by solving the following differential equation

$$\partial_\tau \mathbf{h}(\tau) = -(\boldsymbol{\nu}_\mathbf{x}(\mathbf{h}(\tau),\tau))^{-1}\boldsymbol{\nu}_\tau(\mathbf{h}(\tau),\tau)\,, \tag{5}$$

with initial condition $\mathbf{h}(\tau_0) = \mathbf{x}_0$.

5.3 Basic Tracking Algorithm

Regardless of the decision whether to use the method based on the feature flow field or the implicit function theorem in both cases a streamline has to be traced for all critical points detected in the original vector field. The methods differ only in that in the first case a four-dimensional streamline integration has to be performed whereas for the second method an integration in three dimensions is sufficient.

The basic algorithm then is performed as follows. We start by extracting all critical points on the finest scale level, i.e. the original dataset. Note that this does not guarantee a complete seed set for the topology of the scale-space representation of the data, but is sufficient to compute the scale-space evolution of the critical points under consideration. However, finding all critical points in a sampled vector field is a general problem in topological analysis.

Then for each critical point a streamline is traced based on the evaluation of either equation (3) or (5). Note also that in contrast to the more general case of streamline-based feature tracking in time-dependent flow fields investigated by Theisel and Seidel [21], it is sufficient to do only a forward integration of the streamline here.

In both cases scale-space derivatives have to be computed. This can be either accomplished by precomputing the field $\boldsymbol{\nu}$ for a fixed number of scales or by concurrent filtering/diffusion and stream line integration. The first is easy to implement but involves a high memory usage overhead for storing many three-dimensional vector fields. Since streamline integration is only forward in scale, only two scale levels are necessary when the computation of the scale-space representation and the stream line integration is done concurrently.

Furthermore, because we are working in a linear scale-space we have explicit knowledge about the derivative with respect to τ. Since $\partial_\tau \boldsymbol{\nu}(\mathbf{x}, \tau) = \frac{1}{2}\Delta\boldsymbol{\nu}(\mathbf{x}, \tau)$ the derivative with respect to scale can be derived from the second derivatives of $\boldsymbol{\nu}(\mathbf{x}, \tau)$ in space. Thus both $\partial_\mathbf{x}\boldsymbol{\nu}(\mathbf{x}, \tau)$ and $\partial_\tau\boldsymbol{\nu}(\mathbf{x}, \tau)$ can be computed by differentiation in the spatial dimension followed by a linear interpolation between two consecutive scale levels.

Last, the described streamline integration process has to be stopped as soon as an annihilation event is reached, as will be discussed in the next section.

5.4 Handling Bifurcations

Obviously, equations (3) and (5) are only valid as long as the Jacobian $\boldsymbol{\nu}_\mathbf{x}(\mathbf{x}, \tau)$ is not singular. Therefore we have to monitor the Jacobian matrix during the integration of the scale-space trace in order to capture the bifurcation events and stop the integration of the trace, accordingly. Note that the Jacobian $\boldsymbol{\nu}_\mathbf{x}(\mathbf{x}, \tau)$ is a smooth function of the scale parameter τ and, accordingly, its determinant is also a smooth function of τ. Thus, a sign change of the Jacobian determinant signifies a bifurcation. There are now two possibilities for the kind of event that has occurred. The critical point is either passing through the unstable state of a Hopf-type bifurcation or a fold bifurcation point has been reached and the traced critical point is annihilated. In the second case, the condition is sufficient to detect the event since in an annihilation event always two critical points with opposite sign of their Jacobian determinant are involved. But although a wide range of possible Hopf-type bifurcations can be also detected that way, some of them are hard to detect since the Jacobian determinant does not change its sign but instead passes through a second order zero of the characteristic polynomial. In this case it is very unlikely, that the event would be even noticed. Nevertheless, this problem could be solved by computing the eigenvalues of the Jacobian. However, this poses a large computational overhead that is in fact completely unnecessary. Since we are interested in classifying the critical points according to their scale-space lifetime Hopf bifurcations can be safely neglected because only the type of the critical point is changed not the fact of its existence. That means, the problem is not whether a Hopf bifurcation has been missed, but in the case a zero determinant is encountered if it is an annihilation event or not. Therefore, if $det(\boldsymbol{\nu}_{\mathbf{x}}(\mathbf{x}, \tau))$ is zero we have to check if it is a real zero crossing by going one step further in the direction of the streamline before terminating the streamline integration.

5.5 Predictor-Corrector Tracking

Unfortunately, both methods turned out to be numerically unreliable. Although, they work well for simple generated test datasets both fail to capture the correct behavior for noisy real world datasets. Even when we use a fourth order Runge-Kutta integration scheme with adaptive step size for the computation of the streamlines, the computed traces did very soon deviate from the actual path of the critical points in scale-space that have been computed for comparison by repeated extraction of critical points on multiple scale levels. The major problem seems to be the accuracy of computed derivatives. They have to be numerically approximated from the sampled vector field and, although we use a tricubic interpolation scheme, their accuracy, in particular of the second derivatives, is not very high. Since using even higher order interpolation would give rise to a disproportionately large computational effort this is not an option. Therefore, we have combined the scale-space tracing with a Newton-Raphson method in a predictor-corrector approach. In each step of the streamline integration a prediction is computed by integrating in the direction indicated by either the feature flow field $\mathbf{F}(\mathbf{x}, \tau)$ or the right hand side of equation (5). Afterwards, this prediction is refined using a typically very small number of Newton-Raphson iterations. Now it turned out that even a simple third order Runge-Kutta method is sufficient to compute reliable scale-space traces.

6 Results

In this section we will present results of the above described scale-space tracking approach applied both to a generated test dataset as well as to a real flow dataset.

Our first dataset is an artificial test dataset. It was created by resampling a random generated 10^3 vector field to a 50^3 grid using tricubic filtering. In this dataset 16 first-order critical points (3 saddles, 3 focus saddles and 10 foci) have been detected. The original 16 critical points in combination with streamlines seeded in their vicinity are shown in Fig. 1a. The same dataset after adding some noise is shown in Fig. 1b. In this case, normal-distributed noise was added to 20% of the vector components of the field, which leads to a rather low signal-to-noise ratio of approximately 11dB for this dataset.

The number of critical points that can be detected now is 1307 and the topology is much too complex to be of any practical use. Applying the proposed scale-space tracking scheme, enables us to distinguish between critical points that have been solely introduced due to noise and critical points that represent the dominating flow behavior.

For this test, the points were traced using the scheme derived in Sec. 5.2 over the scales from $\tau = 0$ to $\tau = 1.5$. The computation took approximately 70 second on a machine equipped with an AMD Opteron 2.0GHz processor and 8GB of main memory. 8.5 seconds were spent for computing the scale-space representation and 61 second for the actual streamline integration. The numbers for the feature flow field based method of Sec 5.1 are comparable and identical results are produced for both methods. Figure 2a shows the result of this computation. A gray ramp is applied to indicate the scale-space lifetime of the critical points. Points shown in dark gray are very short lived while light colored points could be traced over the whole scale interval. The actual scale-space traces of the critical points are not shown in the images, since

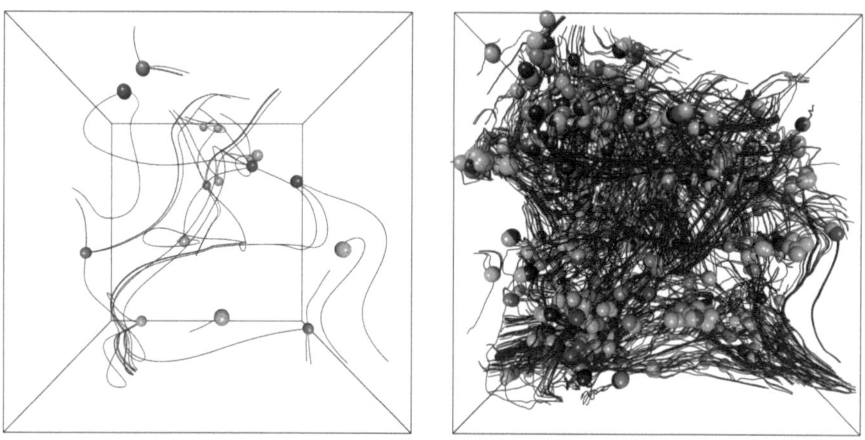

(a) (b)

Fig. 1. A random generated test dataset. (**a**) Original data. (**b**) Data with noise added (colorplate on p. 204).

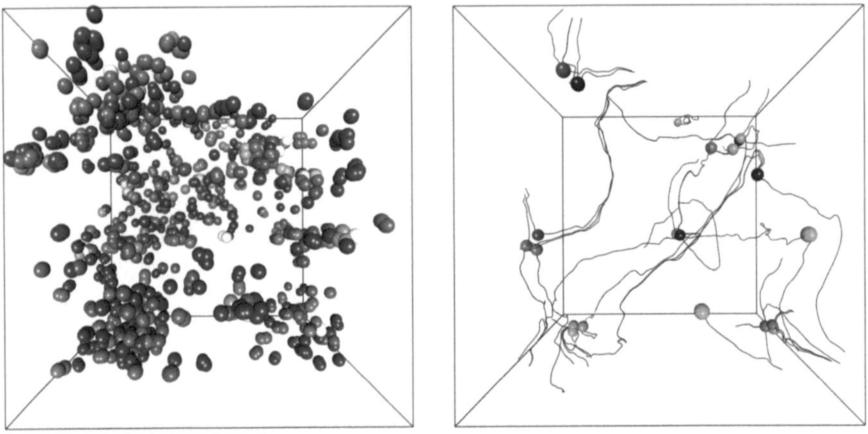

(a) (b)

Fig. 2. The same test dataset as shown in Fig. 1. (**a**) Scale-space lifetime of the critical points in the interval $\tau = 0 \ldots 1.5$ computed by our algorithm. Bright color indicates stable critical points, while dark colored points are very short-lived. (**b**) Critical points filtered by their lifetime. Only points that persist at $\tau_{\min} = 1.0$ are shown (colorplate on p. 204).

they do not provide much of additional information. They may only serve to identify pairs of critical points that participate in annihilation events.

Last, in Fig. 2b the result of filtering the points according to their scale-space lifetime is shown. Only points that have a lifetime larger than $\tau_{\min} = 1.0$ are shown. Note that the streamlines have been integrated starting from the remaining points in the original noisy field not in a smoothed representation. The same holds for the classification of the critical points. Of course, it is not possible to recover the original topology of the noise-free dataset but the overall behavior is much more apparent.

As a second example we present the application of the scale-space tracking to a real CFD dataset – a simulation of the flow past a circular cylinder. There are 141 critical points that can be detected in this dataset, which account for the complex flow topology in the wake behind the cylinder. Figure 3a shows the critical points detected in the flow field. Only the lower half of the 180^3 data set is shown in this image since there were no critical points detected in the upper part of the dataset. For both images of Fig. 3 the same color scheme as for those in Fig. 1 has been used. Computing the scale-space lifetime of the critical points for this dataset took approximately 4:58 minutes using the same machine as for the first example. Since this dataset is significantly larger than the artificial one used in the previous example and also contains far less critical points, most of the time (4:34 minutes) was spent for computing the scale-space flow field representation while for the relatively small number of critical points the scale-space tracking could be

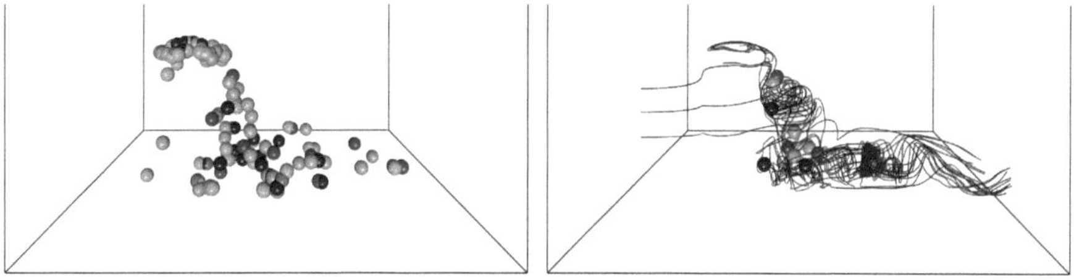

(a) (b)

Fig. 3. The flow behind a spherical cylinder. (**a**) Critical points on the finest scale, i.e. the original flow field. (**b**) Scale-space lifetime of the critical points in the interval $\tau = 0 \ldots 10$ is computed by our algorithm. Critical points were filtered by their lifetime ($\tau_{\min} = 6.3$) and the remaining points have been used to seed stream lines in their vicinity.

done in only 24 seconds. Seeding streamlines only around critical points that can be tracked for $\tau \geq 6.3$ does significantly reduce the amount of streamlines that are displayed and accordingly the problems with visual clutter, but yet one can still clearly discern the overall behavior of the flow. Again, both tracking approaches produce comparable results.

7 Conclusion and Future Work

In this paper we have presented a first approach on a tracking algorithm for vector field singularities in scale-space that uses explicit knowledge of the evolution of the field along the scale-axis. Our approach is based on streamline integration in a derived vector field that allows us to track the evolution of first-order critical points in linear scale-space. Both a feature flow field or the gradient of the level-set implicitly defined by the scale-space curve of a critical point have been used to define that field. Furthermore, we proposed a predictor-corrector method to deal with numerical problems that arise when computing vector field derivatives in a noisy dataset. Last, results have been shown for both artificial test datasets and real CFD data.

Although, the use of the linear scale-space might not be the final answer for building a scale representation of a flow field, we, nevertheless, think that scale-space techniques are a promising way to deal with noisy and highly complex flow datasets. However, a rigorous analysis of the behavior of critical points of 3D vector fields under Gaussian blurring, similar to the work on scalar fields by Kuijper [11], has still to be done.

In future work we want to extend our current implementation to use unstructured grids and other multi-scale analysis schemes. Especially the use of wavelet based smoothing seems to be a promising way to overcome the inherent problem that the linear scale-space representation does not necessarily imply topology simplification. As other possible extensions, we like to generalize the scale-space tracking algorithm also to unsteady flows, i.e. to track the singularities not only over scale but simultaneously over time. This involves the problem of how to compute and interpret the time-surfaces of critical curves in a five-dimensional space. However, this problem is quite similar to the problem of tracking of other extended vector field features, for example vortices. Another general issue in topological analysis based on critical points is the conceptual problem of critical points to be not Galileian-invariant. That means important features can be missed. Therefore we want to look into recent approaches for the treatment of the Galileian-invariance problem [16, 27] in order to guarantee a complete seed point set for the tracking.

Acknowledgments

This work was funded by the German Research Council (DFG) as part of project SFB 382.

References

1. D. Bauer and R. Peikert. Vortex Tracking in Scale-Space. In *Proceedings of the Symposium on Data Visualisation '02*, pages 233–240, 2002.
2. W. de Leeuw and R. van Liere. Visualization of Global Flow Structures Using Multiple Levels of Topology. In *Proceedings of the Symposium on Data Visualisation '99*, pages 45–52, 1999.
3. L. Florack and A. Kuijper. The Topological Structure of Scale-Space Images. *Journal of Mathematical Imaging and Vision*, 12(1):65–79, 2000.
4. C. Garth, X. Tricoche, and G. Scheuermann. Tracking of Vector Field Singularities in Unstructured 3D Time-Dependent Datasets. In *Proceedings of IEEE Visualization '04*, pages 329–336, 2004.
5. A. Globus, C. Levit, and T. Lasinski. A Tool for Visualizing the Topology of Three-Dimensional Vector Fields. In *Proceedings of IEEE Visualization '91*, pages 33–40, 1991.
6. J. Guckenheimer and P. Holmes. *Nonlinear Oscillations, Dynamical Systems, and Bifurcations of Vector Fields*. Springer Verlag, 1986.
7. J. L. Helman and L. Hesselink. Representation and Display of Vector Field Topology in Fluid Flow Data Sets. *IEEE Computer*, 22(8):27–36, 1989.
8. J. L. Helman and L. Hesselink. Visualizing Vector Field Topology in Fluid Flows. *IEEE Comput. Graph. Appl.*, 11(3):36–46, 1991.
9. T. Iijima. Basic theory on normalization of a pattern (in case of typical one-dimensional pattern). *Bulletin of Electrical Laboratory*, 26:368–388, 1962.
10. J. J. Koenderink. The Structure of Images. *Biological Cybernetics*, 50:363–370, 1984.
11. A. Kuijper. *The Deep Structure of Gaussian Scale Space Images*. PhD thesis, Utrecht University, 2002.
12. A. Kuijper and L. Florack. Calculations on Critical Points under Gaussian Blurring. In *Proceedings of the Second International Conference on Scale-Space Theories in Computer Vision '99*, 1999.

13. T. Lindeberg. *Scale-Space Theory in Computer Vision.* Kluwer Academic Publishers, 1994.
14. T. Lindeberg. Edge Detection and Ridge Detection with Automatic Scale Selection. *International Journal of Computer Vision*, 30(2):117–156, 1998.
15. S. Mann and A. Rockwood. Computing Singularities of 3D Vector Fields with Geometric Algebra. In *Proceedings of IEEE Visualization '02*, pages 283–290, 2002.
16. K. Polthier and E. Preuß. Identifying Vector Field Singularities Using a Discrete Hodge Decomposition. In *Visualization and Mathematics III*, pages 113–134. Springer-Verlag, 2003.
17. F. H. Post, B. Vrolijk, H. Hauser, R. S. Laramee, and H. Doleisch. The State of the Art in Flow Visualisation: Feature Extraction and Tracking. *Computer Graphics Forum*, 22(4):775–792, 2003.
18. G. Scheuermann, H. Hagen, H. Krüger, M. Menzel, and A. Rockwood. Visualization of Higher Order Singularities in Vector Fields. In *Proceedings of IEEE Visualization '97*, pages 67–74, 1997.
19. G. Scheuermann, H. Krüger, M. Menzel, and A. P. Rockwood. Visualizing Nonlinear Vector Field Topology. *IEEE Transactions on Visualization and Computer Graphics*, 4(2):109–116, 1998.
20. G. Scheuermann, X. Tricoche, and H. Hagen. C1-Interpolation for Vector Field Topology Visualization. In *Proceedings of IEEE Visualization '99*, 1999.
21. H. Theisel and H.-P. Seidel. Feature flow fields. In *Proceedings of the Symposium on Data Visualisation 2003*, pages 141–148, 2003.
22. H. Theisel, T. Weinkauf, H.-C. Hege, and H.-P. Seidel. Saddle Connectors - An Approach to Visualizing the Topological Skeleton of Complex 3D Vector Fields. In *Proceedings of IEEE Visualization '03*, page 30, 2003.
23. X. Tricoche, G. Scheuermann, and H. Hagen. Continuous Topology Simplification of Planar Vector Fields. In *Proceedings of IEEE Visualization '01*, pages 159–166, 2001.
24. X. Tricoche, T. Wischgoll, G. Scheuermann, and H. Hagen. Topology-Based Visualization of Time-Dependent 2D Vector Fields. In *Proceedings of the Symposium on Data Visualisation '01*, pages 117–126, 2001.
25. X. Tricoche, T. Wischgoll, G. Scheuermann, and H. Hagen. Topology Tracking for the Visualization of Time-Dependent Two-Dimensional Flows. *Computer & Graphics*, 26(2):249–257, 2002.
26. T. Weinkauf, H. Theisel, H.-C. Hege, and H.-P. Seidel. Boundary Switch Connectors for Topological Visualization of Complex 3D Vector Fields. In *Proc. Joint Eurographics - IEEE TCVG Symposium on Visualization (VisSym '04)*, pages 183–192, 2004.
27. A. Wiebel, C. Garth, and G. Scheuermann. Localized Flow Analysis of 2D and 3D Vector Fields. In *Proceedings of Eurographics / IEEE VGTC Symposium on Visualization '05*, pages 143–150, 2005.
28. A. P. Witkin. Scale-Space Filtering. In *Proceedings of the 8th International Joint Conference on Artificial Intelligence*, pages 1019–1022, 1983.

Feature Flow Fields in Out-of-Core Settings

Tino Weinkauf[1], Holger Theisel[2], Hans-Christian Hege[1], and Hans-Peter Seidel[3]

[1] Zuse Institute Berlin (ZIB), Germany
{weinkauf,hege}@zib.de
[2] Bielefeld University, Germany
theisel@techfak.uni-bielefeld.de
[3] MPI Informatik Saarbrücken, Germany
hpseidel@mpi-inf.mpg.de

Summary. Feature Flow Fields (FFF) are an approach to tracking features in a time-dependent vector field $\mathbf{v}$. The main idea is to introduce an appropriate vector field $\mathbf{f}$ in space-time, such that a feature tracking in $\mathbf{v}$ corresponds to a stream line integration in $\mathbf{f}$. The original approach of feature tracking using FFF requested that the complete vector field $\mathbf{v}$ is kept in main memory. Especially for 3D vector fields this may be a serious restriction, since the size of time-dependent vector fields can exceed the main memory of even high-end workstations. We present a modification of the FFF-based tracking approach which works in an out-of-core manner. For an important subclass of all possible FFF-based tracking algorithms we ensure to analyze the data in one sweep while holding only two consecutive time steps in main memory at once. Similar to the original approach, the new modification guarantees the complete feature skeleton to be found. We apply the approach to tracking of critical points in 2D and 3D time-dependent vector fields.

1 Introduction

The resolution of numerical simulations as well as experimental measurements like PIV have evolved significantly in the last years. The challenge of understanding the intricate flow structures within their massive result data sets has made automatic feature extraction schemes popular. Feature-based analysis can be seen as a kind of data reduction since it brings the raw data mass down to a small number of graphical primitives that ought to give insight into the flow structures. While the outcome of most feature extraction algorithms has a rather small memory footprint, the input often exceeds the main memory of high-end workstations. This is especially true for 3D time-dependent data. Thus, feature extraction algorithms should be compatible to an out-of-core data handling, i.e., treating only a small part of the input at once.

A number of algorithms already work in an out-of-core manner. Tricoche et al. [16] and Garth et al. [3] show how to track critical points in piecewise linear vector fields by analyzing the data in one sweep and holding only two time slices at once. Their approaches exploit the linearity of non-changing piecewise linear grids and are probably the best way to go for this important class of data.

Another way of tracking features which are defined by the parallel vector operator [8] is introduced in [1]. This approach is based on a 4-dimensional isosurface extraction – and therefore compatible to out-of-core data handling.

Theisel et al. [13] propose a general approach to feature tracking by capturing the temporal evolution of a feature using a stream object[4] integration in a derived vector field – the feature flow field (FFF). This basic idea has been applied not only to tracking critical points [13] and derived applications like simplification [11] and comparison [12], but also to extracting Galilean invariant vortex core lines [10] and tracking closed stream lines [14]. The FFF approach is independent of an underlying grid, i.e., it is entirely based on the description of the data as a continuous field. At first glance, this concept seems to contradict the principle of out-of-core data handling: treating only a small part of the data at once. In this paper we show that those two concepts do *not* contradict. In fact, we show how all FFF-based tracking algorithms can be formulated in an out-of-core manner. This will be used to re-formulate the algorithm for tracking critical points from [13] to make it compatible to out-of-core data handling. The resulting algorithm enables to analyze the data in one sweep while holding only two time slices at once.

The rest of the paper is organized as follows: section 2 surveys the FFF approach as described in [13] and the basics of out-of-core data handling. Section 3 describes the new out-of-core version of the FFF approach. Section 4 applies this knowledge to FFF-based tracking of critical points, while conclusions are drawn in section 5.

2 Background and Problem

In this section we briefly discuss the basics behind out-of-core data handling and feature flow fields. While their main ideas are not antagonistic, a typical algorithm based on FFF is incompatible with an out-of-core data handling.

2.1 Out-Of-Core Data Handling

Out-of-core refers to the data handling strategy of algorithms, which process data too large to fit into main memory. Thus, only parts of the data can be loaded at once and acted upon. Since loading the data from a mass storage device is very time-consuming, the number of those operations should be reduced to a minimum. This restriction must already be considered when formulating the algorithm.

[4] This refers to a stream line, stream surface, stream volume, etc. – depending on the dimensionality of the feature.

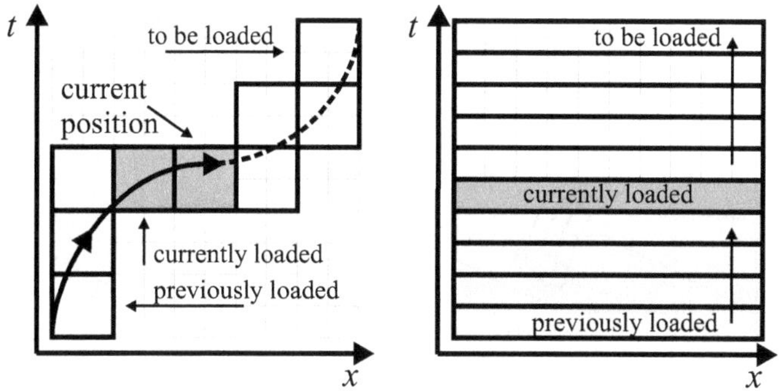

(a) Block-wise random access. (b) Slice-wise sequential access.

Fig. 1. Out-of-core data handling strategies.

There are different types of out-of-core data handling strategies. We just want to mention two here:

- *Block-wise random access:* Data is loaded in blocks of same size. All dimensions are treated equally. The loaded data block with the oldest access time is subject to be substituted with the next block to be loaded. An application for this access pattern is the integration of a path line, which touches only parts of the domain. Figure 1(a) gives an illustration.
- *Slice-wise sequential access:* Data is loaded in slices, i.e., one dimension is fixed for every slice. Slices will be loaded as a fixed sequence in ascending or descending order. The procedure to load all slices from first to last is called a *sweep*. A very common approach is the usage of time slices, since a number of data sets are organized such that each time step is given as a separate file. An application for this access pattern is the extraction of fold bifurcations, where all parts of the domain need to be examined. Figure 1(b) gives an illustration.

Feature extraction algorithms usually do not have a-priori knowledge about the location of the feature and therefore, they need to examine the whole domain. A slice-wise sequential access strategy seems to be even more preferable, if the data is already given in time slices. For the rest of this paper we consider this data handling strategy only. Note, that by loading two consecutive time steps t_i and t_{i+1} and applying a linear interpolation in between them, we obtain the time-dependent field in that time interval.

2.2 Feature Flow Fields

The concept of feature flow fields was first introduced in [13]. It follows a rather generic idea:

Consider an arbitrary point $\mathbf{x}$ known to be part of a feature in a (scalar, vector, tensor) field $\mathbf{v}$. A feature flow field $\mathbf{f}$ is a well-defined vector field at

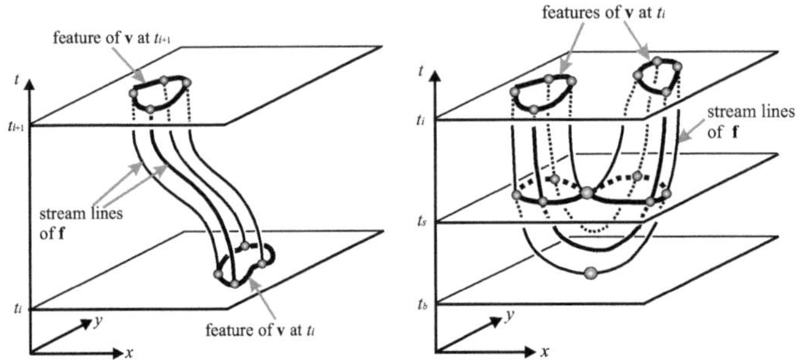

(a) Tracking features by tracing stream lines.

(b) Events: at the time t_b a new feature is born, at t_s it splits into two features.

Fig. 2. Feature tracking using feature flow fields. Features at t_{i+1} can be observed by intersecting these stream lines with the time plane $t = t_{i+1}$.

$\mathbf{x}$ pointing into the direction where the feature continues. Thus, starting a stream line integration of $\mathbf{f}$ at $\mathbf{x}$ yields a curve where all points on this curve are part of the same feature as $\mathbf{x}$.

FFF have been used for a number of applications, but mainly for tracking features in time-dependent fields. Here, $\mathbf{f}$ describes the dynamic behavior of the features of $\mathbf{v}$: for a time-dependent field $\mathbf{v}$ with n spatial dimensions, $\mathbf{f}$ is a vector field $\mathbb{R}^{n+1} \to \mathbb{R}^{n+1}$. The temporal evolution of the features of $\mathbf{v}$ is described by the stream lines of $\mathbf{f}$. In fact, tracking features over time is now carried out by tracing stream lines. The location of a feature at a certain time t_i can be obtained by intersecting the stream lines with the time plane t_i. Figure 2(a) gives an illustration.

Depending on the dimensionality of the feature at a certain time t_i, the feature tracking corresponds to a stream line, stream surface or even higher-dimensional stream object integration. The stream lines of $\mathbf{f}$ can also be used to detect events of the features:

- A birth event occurs at a time t_b, if the feature at this time is only described by one point of a stream object of $\mathbf{f}$, and all stream lines in the neighborhood of this point are in the half-space $t \geq t_b$.
- A split occurs at a time t_s, if one of the stream lines of $\mathbf{f}$ describing the feature touches the plane $t = t_s$ "from above".
- An exit event occurs if all stream lines of $\mathbf{f}$ describing the feature leave the spatial domain.

The conditions for the reverse events (death, merge, entry) can be formulated in a similar way. Figure 2(b) illustrates the different events.

Integrating the stream lines of $\mathbf{f}$ in forward direction does not necessarily mean to move forward in time. In general, those directions are unrelated. The direction in time may even change along the same stream line as it is shown in figure 2(b). This situation is always linked to either a birth and a split event, or a merge and a death event.

Even though we treated the concept of FFF in a rather abstract way, we can already formulate the basics of an algorithm to track all occurrences of a certain feature in a time-dependent field:

Algorithm 1 *General FFF-based tracking*

1. *Get seeding points/lines/structures such that the stream object integration of* $\mathbf{f}$ *guarantees to cover all paths of all features of* $\mathbf{v}$.
2. *From the seeding structures: apply a numerical stream object integration of* $\mathbf{f}$ *in both forward and/or backward direction until it leaves the space-time domain.*
3. *If necessary: remove multiply integrated stream objects.*

Algorithm 1 is more or less an abstract template for a specific FFF-based tracking algorithm like e.g. tracking of critical points. However, it shows a vital contradiction to out-of-core data handling: it gives no guarantee on how the data is processed. We would end up loading different data slices more than once, since both forward and backward integration of $\mathbf{f}$ are allowed, and as already said, the direction in time may even change along the same stream line.

3 Feature Flow Fields and Out-Of-Core Algorithms

In this section we want to modify algorithm 1 such that it becomes compatible to out-of-core data handling. This will allow to formulate all FFF-based algorithms in an out-of-core manner. Before we formulate the algorithm, we collect some concepts and properties of the FFF integration on which the new algorithm is based upon.

3.1 Direction of Integration Regarding Time

The FFF approach is based on a stream line integration of $\mathbf{f}$. Given a starting point $\mathbf{x}_0 = (\mathbf{x}_0^s, t_0)$, $\mathbf{f}$ can be integrated in forward or backward direction. Assuming an Euler integration[5], the forward integration goes to the next point $\mathbf{x}_1 = \mathbf{x}_0 + \varepsilon\, \mathbf{f}(\mathbf{x}_0)$, while the backward integration gives the next point $\mathbf{x}_1 = \mathbf{x}_0 - \varepsilon\, \mathbf{f}(\mathbf{x}_0)$ for a certain small positive ε. In addition to this distinction of the integration orientation, we can also distinguish a t-forward and t-backward orientation. We call an integration *t-forward* if the next point $\mathbf{x}_1 = (x_1^s, t_1)$ is ahead in time, i.e., if $t_1 > t_0$. For $t_1 < t_0$, we have a *t-backward* integration. This property can be decided locally for a point $\mathbf{x}_0$ by looking at the sign of the t-component of $\mathbf{f}(\mathbf{x}_0)$, or if this component is zero, by looking at the sign of the partial derivative $\mathbf{f}_t(\mathbf{x}_0)$. Figure 3 illustrates some of these cases.

[5] For the actual integration we used a fourth order Runge Kutta method, the Euler integration is only for explaining the concepts.

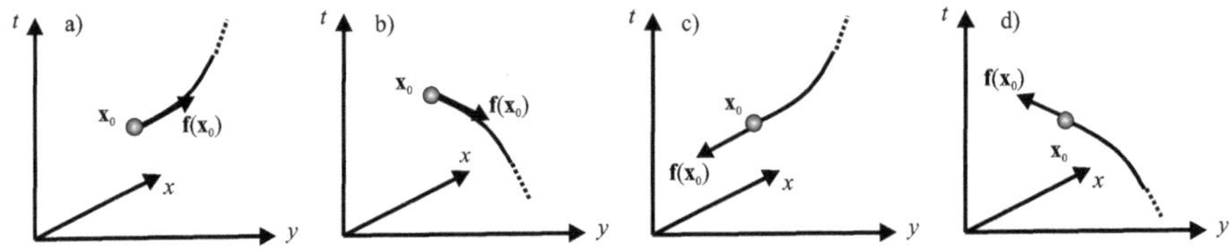

Fig. 3. Orientation of integrating **f**: a) forward and t-forward; b) forward and t-backward; c) backward and t-forward; d) backward and t-backward; the curves are the integrated stream lines starting from $\mathbf{x}_0$.

3.2 Classification of Seeding Structures

The FFF approach is also based on finding appropriate starting structures for the integration. The definition of a complete set of seeding structures is up to the specific FFF-based application. However, we can give the following classification of those structures:

- **t-forward structures:** all integrations started here are t-forward only.
- **t-backward structures:** all integrations started here are t-backward only.
- **intermediate structures:** a t-forward and a t-backward integration will be started here.

This classification is independent of a specific FFF-based application, though it might be that in certain cases a class of structures is empty, e.g. there are only t-forward and intermediate structures but no t-backward structures.

As already discussed in section 2.2, a t-forward integration may change to a t-backward integration even along the same stream line. This situation is always linked to either a birth or a death event, which perfectly fit into the classification: a birth event is a t-forward point, and a death event a t-backward point.

3.3 Out-Of-Core FFF-based Tracking Algorithm

The split of the integration into different directions regarding the time is the conceptual key to an out-of-core version of algorithm 1:

Algorithm 2 *Out-of-core FFF-based tracking*

1. *Load the data in a slice-wise sequential manner from t_{min} to t_{max}. For each time interval between the time slices t_i and t_{i+1}:*
   a) *Get seeding structures such that the stream object integration of* **f** *guarantees to cover all paths of all features of* **v**.
   b) *Classify the seeding structures into t-forward, t-backward and intermediate structures.*
   c) *Start a t-forward integration at all t-forward and intermediate structures. Stop the integration, if*
      i. *the spatial domain was left.*

ii. the temporal domain of the time interval was left, i.e., the integration reached t_{i+1}. Add the result of this integration to the list of t-forward structures for the next time interval.

iii. a death event was reached. If this point will not be reached by any other t-forward integration, add the result to the list of t-backward structures.

2. If the list of t-backward structures is non-empty: load the data in a slice-wise sequential manner from t_{max} to t_{min}. For each time interval between the time slices t_i and t_{i-1} start a t-backward integration at all t-backward and intermediate structures similar as above.

3. Repeat the stream object integrations of steps 1 and 2 until the lists of seeding structures are empty.

The basic idea of this algorithm template is to ensure that the direction of loading the data coincides with the direction of integration, i.e., if we are loading the data from t_{min} to t_{max} then we are integrating t-forward only, and the other way around. This alone does not sound very effective, since the data might been loaded more than once, possibly even an unknown number of times.

This changes, if we take a closer look at the t-forward structures, i.e., all points where *only* a t-forward integration is intended. At those points the features appear for the first time. Examples are entry points, birth events or all occurrences of the feature at t_{min}. If we can find all t-forward structures while doing the first sweep through the data, then the whole feature skeleton can be extracted with this one sweep. This is always fulfilled, if all types of t-forward structures are locally defined, i.e., they can be extracted by a local analysis. Under these prerequisites, we can reformulate algorithm 2 and obtain the following algorithmic template:

Algorithm 3 *One-sweep Out-of-core FFF-based tracking*

1. For each time interval $[t_i, t_{i+1}]$ from t_{min} to t_{max}:

a) Extract all t-forward seeding structures needed to cover all paths of all features of $\mathbf{v}$.

b) Apply a t-forward integration starting at those structures until

i. the spatial domain was left.

ii. the temporal domain of the time interval was left, i.e., the integration reached t_{i+1}. Add the result of this integration to the list of t-forward structures for the next time interval.

iii. a death event was reached.

Algorithm 3 ensures that every path of a feature is integrated only once. Thus, a removal of multiply integrated stream objects is not needed anymore. In comparison to algorithms 1 and 2 it is perfectly fitted for large data sets: it reads only parts of the data and each part only once.

4 Application to Tracking of Critical Points

Critical points, i.e. isolated points with a vanishing flow, are perhaps the most important topological feature of vector fields. For static fields, their extraction and classification is well-understood both in the 2D [6] and the 3D case [17]. Critical points also serve as the starting points of certain separatrices, i.e. stream lines/surfaces which divide the field into areas of different flow behavior. Topological methods have not only been developed for visualization purposes [4, 5], but have also been applied to simplify [2, 15], smooth [18], and compress [11, 7] vector fields. A thorough overview can be found in [9].

Considering a stream line oriented topology of time-dependent vector fields, critical points smoothly change their location and orientation over time. In addition, certain bifurcations of critical points may occur. To capture the topological behavior of time-dependent vector fields, it is necessary to capture the temporal behavior of the critical points.

Theisel et al. introduced in [13] a FFF-based approach to track critical points, which matches algorithm 1. In order to track critical points in an effective out-of-core manner using algorithm 3 we need to find the complete set of t-forward points, i.e., all points in space-time where a critical point appears for the first time. This will be done in section 4.2. But before that we discuss the definition of the feature flow field $\mathbf{f}$ itself.

4.1 FFF for Tracking Critical Points

We first consider tracking critical points in a 2D time-dependent vector field, which is given as

$$\mathbf{v}(x,y,t) = \begin{pmatrix} u(x,y,t) \\ v(x,y,t) \end{pmatrix} \tag{1}$$

in the 3D space-time domain $D = [x_{min}, x_{max}] \times [y_{min}, y_{max}] \times [t_{min}, t_{max}]$. We can construct a 3D vector field $\mathbf{f}$ in D with the following properties: for any two points $\mathbf{x}_0$ and $\mathbf{x}_1$ on a stream line of $\mathbf{f}$, it holds $\mathbf{v}(\mathbf{x}_0) = \mathbf{v}(\mathbf{x}_1)$. This means that a stream line of $\mathbf{f}$ connects locations with the same values of $\mathbf{v}$. Figure 4 gives an illustration. In particular, if $\mathbf{x}_0$ is a critical point in $\mathbf{v}$, then the stream line of $\mathbf{f}$ describes the path of the critical point over time. To get $\mathbf{f}$, we search for the direction in space-time in which both components of $\mathbf{v}$ locally remain constant. This is the direction perpendicular to the gradients of the two components of $\mathbf{v}$. We get

$$\mathbf{f}(x,y,t) = \mathrm{grad}(u) \times \mathrm{grad}(v) = \begin{pmatrix} u_x \\ u_y \\ u_t \end{pmatrix} \times \begin{pmatrix} v_x \\ v_y \\ v_t \end{pmatrix} = \begin{pmatrix} \det(\mathbf{v}_y, \mathbf{v}_t) \\ \det(\mathbf{v}_t, \mathbf{v}_x) \\ \det(\mathbf{v}_x, \mathbf{v}_y) \end{pmatrix}. \tag{2}$$

The FFF approach for 3D vector fields is a straightforward extension of the 2D case. Given the 3D time-dependent vector field

$$\mathbf{v}(x,y,z,t) = \begin{pmatrix} u(x,y,z,t) \\ v(x,y,z,t) \\ w(x,y,z,t) \end{pmatrix} \tag{3}$$

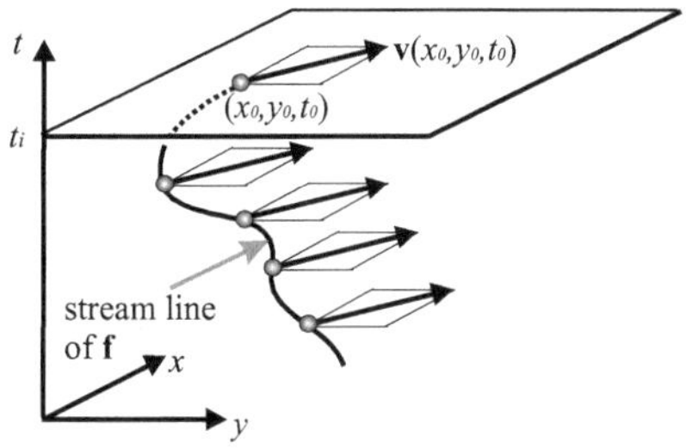

Fig. 4. Tracking 2D critical points: all points on a stream line of **f** have the same value for **v**.

in the 4D space-time domain $D = [x_{min}, x_{max}] \times [y_{min}, y_{max}] \times [z_{min}, z_{max}] \times [t_{min}, t_{max}]$, the 4D FFF **f** is defined by the conditions

$$\mathbf{f} \perp \text{grad}(u) = (u_x, u_y, u_z, u_t)^T \quad , \quad \mathbf{f} \perp \text{grad}(v) \quad , \quad \mathbf{f} \perp \text{grad}(w).$$

This gives a unique solution for **f** (except for scaling)[6]

$$\mathbf{f}(x,y,z,t) = \begin{pmatrix} +\det(\mathbf{v}_y, \mathbf{v}_z, \mathbf{v}_t) \\ -\det(\mathbf{v}_z, \mathbf{v}_t, \mathbf{v}_x) \\ +\det(\mathbf{v}_t, \mathbf{v}_x, \mathbf{v}_y) \\ -\det(\mathbf{v}_x, \mathbf{v}_y, \mathbf{v}_z) \end{pmatrix} . \tag{4}$$

4.2 Complete Set of t-forward Points

Theisel and Weinkauf showed in [14] that two classes of seeding points guarantee that all paths of critical points are captured: the intersections of the paths with the domain boundaries, and fold bifurcations. However, they did not distinguish between t-forward and t-backward points. We are going to do this here in order to find the complete set of t-forward points.

To find all intersections with the boundaries, we have to solve

$\mathbf{v}(x, y, t_{min}) = (0,0)^T$ and $\mathbf{v}(x, y, t_{max}) = (0,0)^T$ for the unknowns x, y,
$\mathbf{v}(x, y_{min}, t) = (0,0)^T$ and $\mathbf{v}(x, y_{max}, t) = (0,0)^T$ for the unknowns x, t,
$\mathbf{v}(x_{min}, y, t) = (0,0)^T$ and $\mathbf{v}(x_{max}, y, t) = (0,0)^T$ for the unknowns y, t.

Each of the 6 solutions turns out to be a simple extraction of critical points of a 2D (steady) vector field. We can make the following distinction:

- *Bottom intersection points* are intersections with the plane $t = t_{min}$
- *Top intersection points* are intersections with the plane $t = t_{max}$
- *Side intersection points* are intersections with the plane $x = x_{min}$, $x = x_{max}$, $y = y_{min}$, or $y = y_{max}$ respectively.

Side intersection points can be further classified into entry and exit points. At an entry point, a t-forward integration of **f** goes into D, while at an exit point

[6] Note that the formulation of $\mathbf{f}(x,y,z,t)$ in [13] contains an error: the alternating signs of the components are missing.

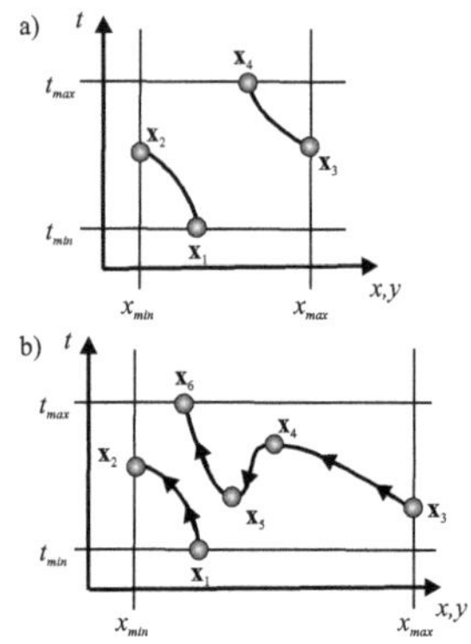

Fig. 5. a) intersections of the paths of critical points with the domain of D: bottom intersection ($\mathbf{x}_1$), exit ($\mathbf{x}_2$) and entry ($\mathbf{x}_3$) side intersection, top intersection ($\mathbf{x}_2$); b) example consisting of a bottom intersection ($\mathbf{x}_1$), exit ($\mathbf{x}_2$) and entry ($\mathbf{x}_3$) side intersection, death ($\mathbf{x}_4$) and birth ($\mathbf{x}_5$) fold bifurcation, top intersection ($\mathbf{x}_6$): the paths of critical points consist of 4 segments which are integrated t-forward from→to: $\mathbf{x}_1 \rightarrow \mathbf{x}_2$, $\mathbf{x}_3 \rightarrow \mathbf{x}_4$, $\mathbf{x}_5 \rightarrow \mathbf{x}_4$, $\mathbf{x}_5 \rightarrow \mathbf{x}_6$.

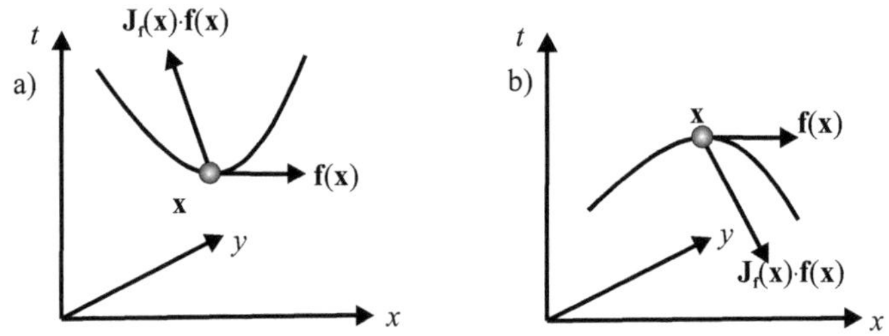

Fig. 6. Classifying fold bifurcations by the last component of $\mathbf{J}_\mathbf{f}(\mathbf{x}) \cdot \mathbf{f}(\mathbf{x})$: a) birth event; b) death event.

a t-forward integration leaves D. Figure 5a illustrates the different kinds of intersection points with the boundary of D. It is easy to see that only bottom and entry side intersections are t-forward points.

To detect fold bifurcations inside D, we search for locations $\mathbf{x}$ with

$$[\ \mathbf{v}(\mathbf{x}) = (0,0)^T \ \ , \ \ \det(\mathbf{J}_\mathbf{v}(\mathbf{x})) = 0\ ] \tag{5}$$

where $\mathbf{J}_\mathbf{v}$ is the Jabcobian matrix of $\mathbf{v}$. (2) shows that the second condition of (5) ensures that the last component of $\mathbf{f}$ vanishes, i.e., that $\mathbf{f}$ is parallel to the t-axis. To solve (5), we use a numerical approach similar to extracting isolated critical points in 3D vector fields. There are two kinds of fold bifurcations: birth and death events. To distinguish them, we consider the last component of $\mathbf{J}_\mathbf{f}(\mathbf{x}) \cdot \mathbf{f}(\mathbf{x})$ at the fold bifurcation. If this component is positive, we have a birth bifurcation; if it is negative, a death bifurcation is present. Figure 6 illustrates this. It is easy to see that only birth bifurcations are t-forward points.

For 3D time-dependent vector fields, the extraction of the seeding points follows the same ideas. Boundary intersections are found as isolated critical points of the 3D (steady) vector fields at the space-time domain boundary. Again, only bottom and entry side intersections are of interest. Fold bifurcations are the solutions of

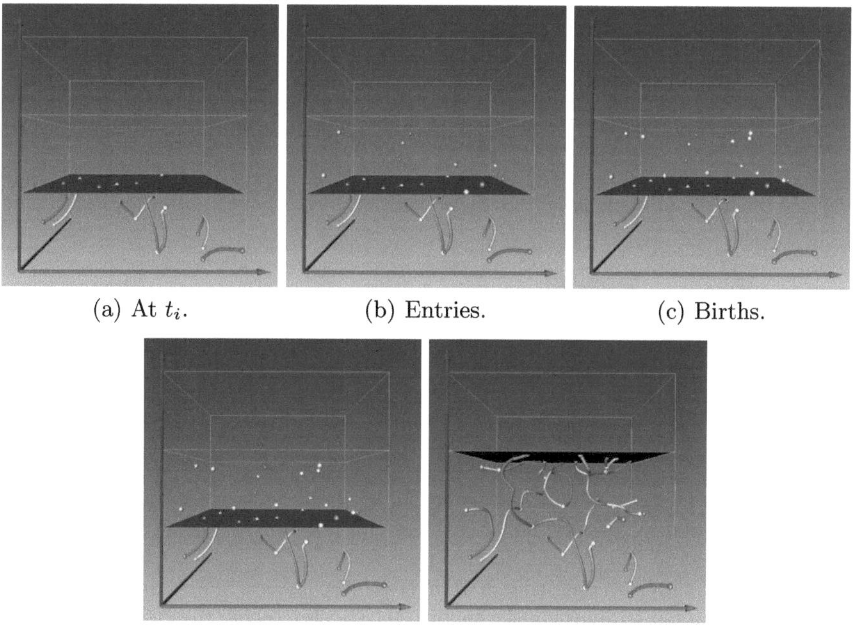

(a) At t_i. (b) Entries. (c) Births.

(d) Integration. (e) At t_{i+1}.

Fig. 7. Application of algorithm 3: critical points tracked in one sweep (colorplate on p. 205).

$$[\ \mathbf{v}(\mathbf{x}) = (0,0,0)^T\ ,\ \ \det(\mathbf{J}_\mathbf{v}(\mathbf{x})) = 0\ ] \tag{6}$$

which corresponds to numerically finding isolated critical points in 4D vector fields. The distinction between births and deaths follows the 2D case.

We have found all points in space-time where a critical point can appear for the first time: bottom and entry side intersections as well as birth bifurcations. They can all be extracted using a local analysis. All prerequisites for algorithm 3 are fulfilled. Thus, we are now able to track critical points in 2D and 3D time-dependent vector fields in an effective out-of-core manner: in one sweep and by loading only two slices at once. We applied this algorithm to a random 2D time-dependent data set. Random vector fields are useful tools for a proof-of-concept of topological methods, since they contain a maximal amount of topological information. Figure 7 shows the execution of algorithm 3 between two consecutive time steps t_i and t_{i+1}.

Figure 8 shows the visualization of a vector field describing the flow over a 2D cavity. This data set was kindly provided by Mo Samimy and Edgar Caraballo (both Ohio State University) as well as Bernd R. Noack and Ivanka Pelivan (both TU Berlin). 1000 time steps have been simulated using the *compressible* Navier-Stokes equations; it exhibits a non-zero divergence inside the cavity, while outside the cavity the flow tends to have a quasi-divergence-free

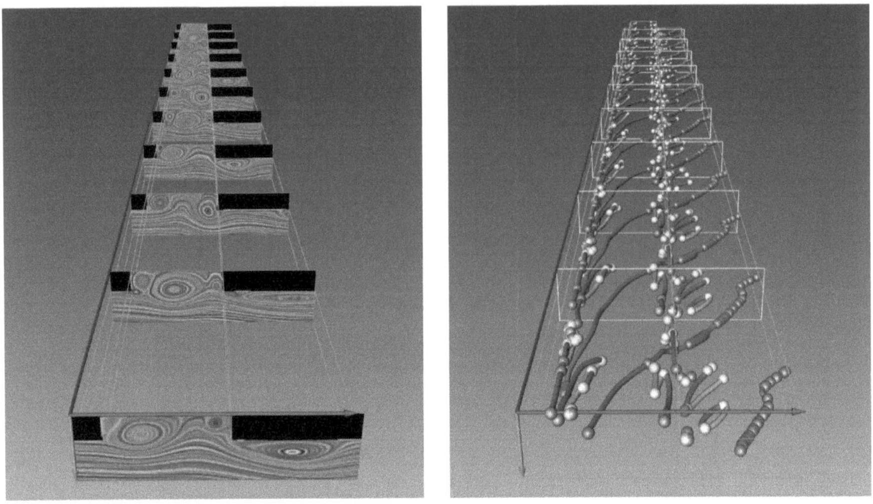

(a) LIC planes: 10 sections. (b) Tracked critical points.

Fig. 8. Cavity data set consisting of 1000 time steps. Algorithm 3 has been applied onto the 10 depicted sections consisting of 100 time steps each. (Colorplate on p. 205)

behavior. Instead of loading only two time steps at once, the data set has been divided into 10 sections each consisting of 100 time steps (Figure 8(a)). Each section fits easily into main memory and has been treated according to algorithm 3. This approach reduced the overhead introduced by the out-of-core handling. The computation time for this data set was 20 minutes. The topological structures visualized in Figure 8(b) elucidate the quasi-periodic nature of the flow. The most dominating topological structures originate in or near the boundaries of the cavity itself. The quasi-divergence-free behavior outside the cavity is affirmed by the fact that a high number of Hopf bifurcations has been found in this area.

5 Conclusions

In this paper we showed how all FFF-based tracking algorithms can be formulated in an out-of-core manner. This has been used to re-formulate the algorithm for tracking critical points from [13] to make it compatible to out-of-core data handling. The resulting algorithm enables to analyze the data in one sweep while holding only two time slices at once. For future work, we intend to apply algorithm 3 to other types of features, especially to vortex core lines.

References

1. D. Bauer and R. Peikert. Vortex tracking in scale space. In *Data Visualization 2002. Proc. VisSym 02*, pages 233–240, 2002.
2. W. de Leeuw and R. van Liere. Collapsing flow topology using area metrics. In *Proc. IEEE Visualization '99*, pages 149–354, 1999.
3. C. Garth, X. Tricoche, and G. Scheuermann. Tracking of vector field singularities in unstructured 3D time-dependent datasets. In *Proc. IEEE Visualization 2004*, pages 329–336, 2004.
4. A. Globus, C. Levit, and T. Lasinski. A tool for visualizing the topology of three-dimensional vector fields. In *Proc. IEEE Visualization '91*, pages 33–40, 1991.
5. H. Hauser and E. Gröller. Thorough insights by enhanced visualization of flow topology. In *9th international symposium on flow visualization*, 2000.
6. J. Helman and L. Hesselink. Representation and display of vector field topology in fluid flow data sets. *IEEE Computer*, 22(8):27–36, 1989.
7. S. Lodha, N. Faaland, and J. Renteria. Topology preserving top-down compression of 2d vector fields using bintree and triangular quadtrees. *IEEE Transactions on Visualization and Computer Graphics*, 9(4):433–442, 2003.
8. R. Peikert and M. Roth. The parallel vectors operator - a vector field visualization primitive. In *Proc. IEEE Visualization 99*, pages 263–270, 1999.
9. F.H. Post, B. Vrolijk, H. Hauser, R.S. Laramee, and H. Doleisch. Feature extraction and visualisation of flow fields. In *Proc. Eurographics 2002, State of the Art Reports*, pages 69–100, 2002.
10. J. Sahner, T. Weinkauf, and H.-C. Hege. Galilean invariant extraction and iconic representation of vortex core lines. In *Proc. EuroVis 2005*, pages 151–160, 2005.
11. H. Theisel, Ch. Rössl, and H.-P. Seidel. Combining topological simplification and topology preserving compression for 2d vector fields. In *Proc. Pacific Graphics 2003*, pages 419–423, 2003.
12. H. Theisel, Ch. Rössl, and H.-P. Seidel. Using feature flow fields for topological comparison of vector fields. In *Proc. Vision, Modeling and Visualization 2003*, pages 521–528, 2003.
13. H. Theisel and H.-P. Seidel. Feature flow fields. In *Data Visualization 2003. Proc. VisSym 03*, pages 141–148, 2003.
14. H. Theisel, T. Weinkauf, H.-C. Hege, and H.-P. Seidel. Topological methods for 2D time-dependent vector fields based on stream lines and path lines. *IEEE Transactions on Visualization and Computer Graphics*, 11(4):383–394, 2005.
15. X. Tricoche, G. Scheuermann, and H. Hagen. Continuous topology simplification of planar vector fields. In *Proc. Visualization 01*, pages 159 – 166, 2001.
16. X. Tricoche, T. Wischgoll, G. Scheuermann, and H. Hagen. Topology tracking for the visualization of time-dependent two-dimensional flows. *Computers & Graphics*, 26:249–257, 2002.
17. T. Weinkauf, H. Theisel, H.-C. Hege, and H.-P. Seidel. Boundary switch connectors for topological visualization of complex 3D vector fields. In *Proc. VisSym 04*, pages 183–192, 2004.
18. R. Westermann, C. Johnson, and T. Ertl. Topology-preserving smoothing of vector fields. *IEEE Transactions on Visualization and Computer Graphics*, 7(3):222–229, 2001.

Streamline Predicates as Flow Topology Generalization

Tobias Salzbrunn and Gerik Scheuermann

University of Leipzig
{salzbrunn,scheuermann}@informatik.uni-leipzig.de

Summary. Streamline predicates are simply boolean functions on the set of all streamlines in a flow field. A characteristic set of a streamline predicate is the set of all streamlines fulfilling the predicate. If streamline predicates are defined based on asymptotic behavior, the characteristic sets become α- or ω-basins. Using boolean algebra on the streamline predicates, we obtain the usual flow topology. We show that these considerations allow us to generalize flow topology to flow structure definitions. These flow structure definitions can be flexibly adapted to typical analysis tasks arising in flow studies and taylored to the users' needs

1 Introduction

Flow topology has been developed into a tool that gives information about the course of streamlines in steady two and three-dimensional velocity vector fields. Basically, it clusters streamlines with similar behavior. The clustering is based on a precise definition, namely the basins of dynamical system theory. Therefore, each cluster can be interpreted clearly by the user. This is one of the advantages of topology compared to other clustering methods, like typical statistical clustering [6], [16], anisotropic diffusion [13] or an algebraic multigrid approach [5].

But it must be said that there are also limitations. One drawback is missing Galilean invariance. Topology changes between a fixed observer and an observer moving with constant velocity (because the streamlines change their course). We think that this problem can be solved in many cases by simply taking the given observer of the data. This is useful in typical flows around a single airplane, train, or other obstacle. It is also an obvious choice for flows inside, e.g., a building, cabin or turbine. In more complex situations, we suggest the use of the localized flow approach of Wiebel et al. [22] that allows to remove any flows crossing the outer boundary and is Galilean invariant without creating flow through solid boundaries like the popular method of removing the average flow.

Another, in our eyes more important, limitation of topology is that it may miss relevant aspects of streamline behavior. The most important example are vortices (areas of high vorticity) in three-dimensional flows. Quite often, topology groups streamlines obviously entering the vortex and streamlines not passing the vortex region into the same group because they belong to the same basin in the sense of dynamical system theory. We give a realistic example in the results section. Similar problems can arise with streamlines crossing shocks or entering shear flow areas. Since engineers and scientists often like to distinguish streamlines entering and not entering a vortex, we suggest a solution in this paper that refines topology in these cases.

There is a further limitation of steady flow topology that can hinder understanding: topology does not depend on absolute velocity. Topology concentrates on the set of points visited by a streamline but the visit time does not play a role. But sometimes, engineers are interested only in fast dynamics or the time a particle resides near a surface. We will show that these concepts can be easily expressed by streamline predicates and could therefore be used to enrich topology.

2 Related Work

Of course, this paper builds on quite a large number of publications in flow topology, especially [7, 9, 14, 15, 23, 18, 10, 17, 21, 3]. Besides, we have been also influenced by feature-based visualization [12], especially the early work of van Walsum et al. [19], the work on vortex detection by Peikert et al. [11, 1] and the feature definition language of Doleisch and Hauser [2]. In the previous section, we have already mentioned relevant articles on cluster-based flow visualization.

3 Streamline Predicates

We concentrate our consideration on steady three-dimensional flows. Of course, the planar case is quite similar. Let $D \subset R^3$ be the ***domain***. A ***vector field*** on D is a Lipschitz continuous map

$$\begin{aligned} v : D &\to R^3, \\ x &\mapsto v(x). \end{aligned}$$

A ***streamline of v passing through the point*** $a \in \mathbf{D}$ is a continuous map

$$s_a : J_a \to D$$

where $0 \in J_a \subset R$ is an interval of maximal extend and s_a fulfills the conditions

$$s_a(0) = a,$$
$$\dot{s_a}(\tau) = v(s_a(\tau)) \ \ \forall \tau \in J_a.$$

Since we are interested in the set of streamlines, we identify two streamlines $s_a \simeq s_b$ if there is a $\tau_0 \in R$ such that

$$s_a(\tau) = s_b(\tau + \tau_0)$$

and note that $s_a \simeq s_b$ if there are $\tau \in J_a$, $\tau' \in J_b$ with $s_a(\tau) = s_b(\tau')$ due to the existence and uniqueness theorem for streamlines. We define the set of all streamlines as the set $\mathcal{S}$ of all equivalence classes.

Let $s_\lambda : J_\lambda \rightarrow D$ be a representive of $S_\lambda \in \mathcal{S}$. Every other representive could then be written as $s'_\lambda : J_\lambda + \tau_0 \rightarrow D$, $s'_\lambda(\tau + \tau_0) = s_\lambda(\tau)$. Since the set of points $s_\lambda(J_\lambda)$ (the course of the streamline) is the same for equivalent streamlines, we can define it as $S_\lambda(J_\lambda) := s_\lambda(J_\lambda)$. Then, we have a partition of $D = \bigcup_{S_\lambda \in \mathcal{S}} S_\lambda(J_\lambda)$, since the equivalence classes are mutually disjoint.

A ***streamline predicate*** is defined as a map

$$SP: \ \mathcal{S} \rightarrow \{ \ TRUE, FALSE \ \},$$
$$S \mapsto SP(S).$$

i.e. a boolean map on the streamlines that does not depend on the absolute time at only one position. It may nevertheless depend on relative time between different positions along the streamline.

The ***characteristic set*** of a streamline predicate is defined as

$$C_{SP} := \bigcup_{S_\lambda \in S, \ SP(S) = TRUE} S_\lambda(J_\lambda) \subset D.$$

4 Flow Structure

Our goal in this paper is a definition of flow structure that meets the needs of users in all cases and extends flow topology. A flow structure is considered a partition of the flow into disjunct clusters. We suggest a partition based on streamlines. This agrees with the approach taken by flow topology. Since we want to have a general grouping mechanism, we start with a finite set $\mathcal{G}$ of streamline predicates

$$\mathcal{G} = \{ \ SP_\lambda \mid \lambda \in \Gamma \ \}$$

which is chosen such that their characteristic sets are disjoint, i.e.

$$C_{SP_\lambda} \cap C_{SP_\mu} = \emptyset \ \ \forall \lambda, \mu \in \Gamma.$$

As ***flow structure***, we define the partition of

$$\mathcal{S} = \bigcup C_{SP_\lambda}$$

where we assume that every streamline fulfills exactly one streamline predicate. If $\mathcal{G}$ creates only a partial partition of $\mathcal{S}$, we add the predicate

$$\begin{aligned} SP_0 : \ & \mathcal{S} \to \{ \ TRUE, FALSE \ \}, \\ & S \mapsto \bigwedge_{\lambda \in \Gamma} \{ \ SP_\lambda(S) = FALSE \ \}. \end{aligned}$$

In the next section we show that the usual flow topology is a special flow structure.

5 Flow Topology as Flow Structure

Following Scheuermann et al. [15], we define topology using α- and ω-limit sets. For a streamline s, we define its α***-limit set*** $A(s)$ as

$$A(s) := \{ \ p \in R^3 \mid \exists (t_n)_{n=0}^{\infty} \subset R, \ t_n \to -\infty, \ \lim_{n \to \infty} s(t_n) = p \ \}$$

and its ω***-limit set*** as

$$\Omega(s) := \{ \ p \in R^3 \mid \exists (t_n)_{n=0}^{\infty} \subset R, \ t_n \to \infty, \ \lim_{n \to \infty} s(t_n) = p \ \}.$$

If a streamline enters or leaves the domain D at the boundary ∂D, we define the boundary ∂D as α- resp. ω-limit set.

The union of all streamlines with α-limit set A is called the α***-basin of*** $\boldsymbol{A}$

$$B_\alpha(A) = \{ \ a \in D \mid A(s_a) = A \ \}.$$

Similarly, the union of all streamlines with ω-limit set Ω is called the ω***-basin of*** Ω

$$B_\omega(\Omega) = \{ \ a \in D \mid \Omega(s_a) = \Omega \ \}.$$

If $A_i, \ i \in I$, and $\Omega_j, \ j \in J$, denote all α- and ω-limit sets in D and $Z_k(M)$ denotes the connected components of $M \subset D$, the ***flow topology of*** $\boldsymbol{v}$ can be described as the partition

$$D = \bigcup_{i,j,k} Z_k(B_\alpha(A_i) \cap B_\omega(\Omega_j)).$$

In our framework, we use the following predicates

$$\begin{aligned} SP_{A_i} : \ & \mathcal{S} \to \{ \ TRUE, FALSE \ \}, \ i \in I, \\ & S \mapsto A(S) = A_i. \end{aligned}$$

$$\begin{aligned} SP_{\Omega_j} : \mathcal{S} &\to \{ \, TRUE, FALSE \, \}, \; j \in J, \\ S &\mapsto \Omega(S) = \Omega_j \end{aligned}$$

since α- and ω-basins are the same for equivalent streamlines. We get the basins of the topology as characteristic sets, i.e.

$$C_{SP_{A_i}} = B_\alpha(A_i) \quad C_{SP_{\Omega_i}} = B_\omega(\Omega_i).$$

Therefore, we can use the set of predicates

$$\mathcal{G}_{TOP} = \{ \; SP_{A_i} \; AND \; SP_{\Omega_j} \mid i \in I, \; j \in J\},$$

as definition of a flow structure that coincides with flow topology.

6 Refinement of Flow Topology

Looking at section 5, we can ask what is gained by using streamline predicates and general flow structures compared to flow topology. The answer is a wide flexibility because there is no reason to choose exactly the predicates used in the previous section.

We want to show this flexibility using an important example in practice. A user studies steady flow around an obstacle (car, airplane, train, sphere, ellipsoid, house, ...) and is interested in vortices. For the flow topology, critical points, closed streamlines, and boundary switch points are determined. As next step, separating surfaces and isolated streamlines starting at saddle points are computed. Including an analysis of the boundary of the obstacle, it is likely that even for vortices close to a typical model like Vatistas [20], the vortex will show up only as a single streamline. Streamlines obviously rotating around this line and streamlines not rotating around the line will be in the same topological component. At this point, streamline predicates can show their strength. In a first step, the user can apply any vortex detection method, e.g. the λ_2-method of Jeong and Hussein [8], and define the extend of vortices. In a second step, he defines a streamline predicate for each vortex that decides if the streamline crosses the vortex region. The third step creates a flow structure using all and-combinations of the SP_{A_i} *and* SP_{Ω_j} of predicates from topology with the vortex predicates and their opposite predicates. In this way, streamlines entering the vortex are distinguished from streamlines missing it.

Of course, whenever the user defines interests in the behavior of streamlines with streamline predicates, a similar solution is possible. Therefore, flow structure based on streamline predicates allows a refinement of flow topology tailored to the users needs.

7 Results

In the remainder of this paper we want to present three examples of streamline predicates addressing questions which could not be answered with flow topology methods. The dataset we use corresponds to a single time step of an unsteady simulation of the German train ICE. The train travels at a velocity of about 250 km/h with a wind blowing from the side at an angle of 30 degrees. The wind causes vortices to form on the lee side of the train, creating a drop in pressure that has adverse effects on the trains track holding. For our computations we choose a region of interest around the front wagon. To represent the set of all streamlines $\mathcal{S}$ we choose a finite subset $\tilde{\mathcal{S}}$. We use a Cartesian grid in the area $[-15000, 45000] \times [-15000, 25000] \times [350, 5500]$ with 200 units as spacing in all directions as starting positions for the streamlines in $\tilde{\mathcal{S}}$. This is a set of more than 1.56 million streamlines that fills the space around the train in a dense manner.

The first predicate is exemplary for streamline predicates using time information of steady vector fields (i.e. absolute velocity). We are interested in parts of the flow which have a direct influence on the surface (and immediate neighborhood) of the train. Especially particles residing a "long" time near the surface are of interest. Of course one could use a fixed minimum residence time given by some physical considerations for a given application area. However we take another approach and calculate the residence time for a representative set of streamlines to get an idea of a meaningful value. From the resulting distribution we take the value of the 99-% quantile as minimum residence time t_{min}. For the required minimum distance calculation we compute a distance field on the positions of the dataset grid thus reducing minimum distance calculations to a simple interpolation in the distance field at a questioned position. Fig. 1 shows the isosurface of the distance field for an isovalue of 20 [cm] (which we use as maximal neighborhood distance for our computations). We define the following general streamline predicate (instantiated with the previous values):

$$A - \tilde{\mathcal{S}} \ stays \ a \ minimum \ time \ t_{min} \ in \ the \ neighborhood \ of \ an \ object$$

The resulting flow structure $\mathcal{G}_{Surface} = \{ A, \bar{A} \}$ is of course very simple, but will get more complex if more than one object is taken into account. Fig. 1 shows the boundary of the resulting characteristic set A. There is one part of the flow hitting the train on the luv side and flowing around the train and a second part hitting the head of the train and being pushed towards the trains surface (lee side).

In the second example we want to study the deviation of the flow from the principal input flow direction thus getting the most turbulent parts of the flow. To compute the deviation we integrate the difference between the tangent vector direction and the main inflow direction along the streamlines. Again we sample a representative set of streamlines, compute the deviation and take the 99-% quantile as minimum deviation d_{min}. We define the streamline predicate:

$$D \quad - \textit{Deviation of} \tilde{S} \textit{ from principal direction is greater than } d_{min}$$

Fig. 2 shows the boundary of the resulting characterist set D. One can see the flow that deviates very strong from the principal inflow direction.

In the third example we examine the interplay between vortices in the flow and the flow regions outside the vortex regions. Applicability of topological methods is limited concerning this important application domain due to the lack of singularities of the velocity field. We examine if a streamline enters a certain vortex region in order to test if a streamline is influenced by a vortex. Of course more precise and sophisticated methods are possible, but we decide to hold computational effort down. To compute the vortex regions we use the λ_2-criteria of Jeong and Hussain [8]. The λ_2-criteria does not clearly separate vortex regions of different vortices, especially if they are close together. To address this issue we compute as additional information about the vortices the vortex core lines with the gravity-line-method explained in [4]. The resulting vortex core lines are depicted in Fig. 3. Based upon the vortex core lines we use a flood-fill algorithm to label each cell according to which vortex region (if any) it belongs to. We start with the cells that inherit a segment of a vortex core line computed in the previous step. Each cell is examined for its λ_2-value. To get a cell based λ_2-value we assign to every cell the mean of the λ_2-values of its vertices. If the cell has a negative λ_2-value it gets the label of the respective vortex core. The labeled cells are put into a priority queue with the most negative λ_2-value on top. After the initial feeding of the priority queue the neighboring cells of the top element of the queue are examined for their λ_2-values. If a neighboring cell with negative λ_2-value exists it gets the label of the top element. The top element is removed afterwards and the labeled neighboring cells are inserted according to their λ_2-values. This strategy insures that regions with strong vortices grow faster. The flood-fill algorithm is finished if the queue is empty. We now have vortex cores with corresponding regions as a set of cells with the appropriate label. The vortex regions of the train-dataset are depicted in Fig. 3. To compute the following streamline predicates one has to check if a streamline enters the cells of a vortex core.

We evaluate the three streamline predicates

$$R \quad - \tilde{S} \textit{ enters the red vortex region}$$
$$G \quad - \tilde{S} \textit{ enters the green vortex region}$$
$$B \quad - \tilde{S} \textit{ enters the blue vortex region}$$

For the flow structure, we need a set of streamline predicates with disjunct characteristic sets filling up D. Unfortunately, the lack of singularities prohibited to compute the SP_{A_i} *and* SP_{Ω_j} predicates from topology. Additionally attempts to start from the surface topology were not successful. We did not examine boundary switch connectors as proposed in [21], but we assume that they will not separate the vortices in a way one would expect it according to the λ_2-criteria. Therefore, we choose the set

$$\begin{aligned}\mathcal{G}_{Vortex} = \{\ & \bar{R}\wedge\bar{G}\wedge\bar{B},\ \ \bar{R}\wedge\bar{G}\wedge B,\ \ \bar{R}\wedge G\wedge\bar{B},\\ & \bar{R}\wedge G\wedge B,\ \ R\wedge\bar{G}\wedge\bar{B},\ \ R\wedge\bar{G}\wedge B,\\ & R\wedge G\wedge\bar{B}, R\wedge G\wedge B\ \},\end{aligned}$$

In this way, we separate streamlines by the vortex regions they enter. $\bar{R}\wedge G\wedge B$, for example describes the streamlines entering the green and the blue vortex region, but not entering the red vortex region. Fig. 4 and 5 show the boundaries of all characteristic sets of $\mathcal{G}_{Vortex}$ (except $\bar{R}\wedge\bar{G}\wedge\bar{B}$).

8 Conclusion

We introduced streamline predicates as a new tool to study flow datasets. We showed that a flow structure based upon appropriate streamline predicates comprises and refines flow topology. Applied to one realistic CFD-dataset, streamline predicates proofed able to answer questions where conventional topological methods could not be applied. Computing the streamlines and the characteristic sets requires high computational effort for brute force implementations. Further research should deal with increasing the efficiency of the computations.

Acknowledgment

The authors wish to thank Markus Rütten, from German Aerospace Center(DLR) in Göttingen for providing the datasets. We also wish to thank Xavier Tricoche and Christoph Garth for their valuable advice. Last but not least thanks go to all members of the FAnToM development team for their programming efforts. This work was partly supported by DFG grants HA 1491/15-5 and SCHE 663/3-7.

References

1. D. Bauer and Peikert R. Vortex Tracking in Scale Space. In D. Bauer, P. Brunet, and I. Navazo, editors, *Proceedings of Eurographics/IEEE-VGTC Symposium on Visualization 2002 (EuroVis 2002)*, pages 233–240, 2002.
2. H. Doleisch, M. Gasser, and H. Hauser. Interactive Feature Specification for Focus+Context Visualization of Complex Simulation Data. In *Proceedings of the 5th Joint IEEE TCVG - EUROGRAPHICS Symposium on Visualization (VisSym 2003)*, pages 239 – 248, 2003.
3. C. Garth, X. Tricoche, and G. Scheuermann. Tracking of vectorfield singularities in unstructured 3d-time dependent datasets. In *IEEE Visualization 2004*, pages 329 – 336, Austin, Texas, 2004.
4. C. Garth, Tricoche X., T. Salzbrunn, Bobach T., and G. Scheuermann. Surface Techniques for Vortex Visualization. In *VisSym*, pages 155–164, 346, 2004.

Fig. 1. Boundary of the characteristic set A of the flow structure $\mathcal{G}_{Surface}$. (Upper left picture shows the hull around the train.)

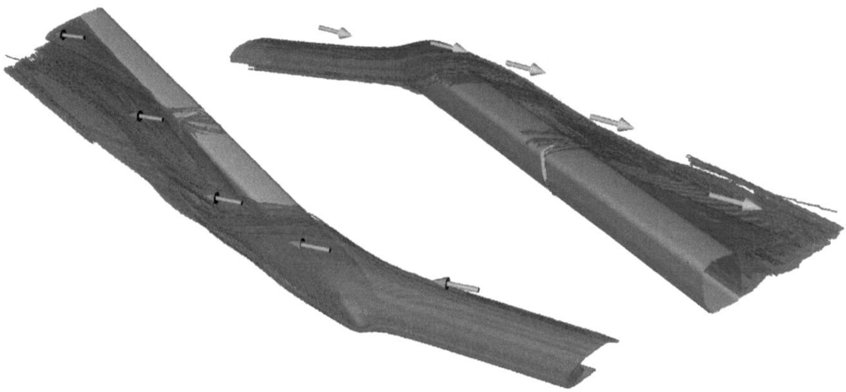

Fig. 2. Boundary of the characteristic set D of the deviation predicate. (The arrows show the principal direction of the entry flow.)

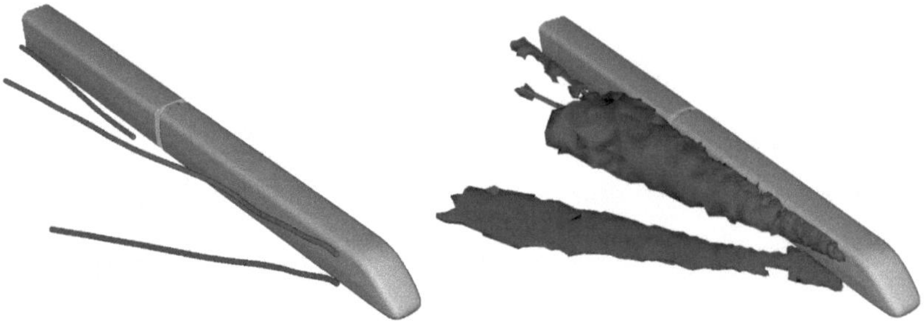

Fig. 3. The left picture shows the vortex core lines of the train-dataset. In the right picture the corresponding vortex core regions are depicted. (Note that the vortex core regions do not cover the vortex core lines completely. We assume that the vortices level off at the end.) (Colorplate on p. 206.)

5. M. Griebel, T. Preusser, M. Rumpf, M.A. Schweitzer, and A. Telea. Flow Field Clustering via Algebraic Multigrid. In *IEEE Visualization 2004*, pages 35 – 42, Austin, Texas, 2004.
6. B Heckel, G.H. Weber, B Hamann, and K.I. Joy. Construction of Vector Field Hierarchies. In *IEEE Visualization 1999*, pages 19–25, San Francisco, CA, 1999.
7. J. L. Helman and L. Hesselink. Visualizing Vector Field Topology in Fluid Flows. *IEEE Computer Graphics and Applications*, 11(3):36–46, May 1991.
8. J. Jeong and F. Hussain. On the Identification of a Vortex. *Journal of Fluid Mechanics*, 285:69 – 94, 1995.
9. H. Löffelmann, T Kucera, and Gröller E. Visualizing Poincare Maps Together with the Underlying Flow. In H.C. Hege and K. Polthier, editors, *Proceedings of the International Workshop on Visualization and Mathematics 1997 (Vis-Math'97)*, pages 315–328, 1998.
10. K. Mahrous, J. Bennett, G. Scheuermann, B. Hamann, and K. I. Joy. Topological Segmentation of Three-Dimensional Vector Fields. *IEEE Transactions on Visualization and Computer Graphics*, 10(2):198 – 205, 2004.
11. R. Peikert and M. Roth. The Parallel Vectors Operator - a Vector Field Visualization Primitive. In *IEEE Visualization 1999*, pages 263 – 270, San Francisco, CA, 1999.
12. F.H. Post, B. Vrolijk, H. Hauser, R.S. Laramee, and H. Doleisch. The State of the Art in Flow Visualization: Feature Extraction and Tracking. In *Computer Graphics Forum 22*, volume 4, pages 775–792, 2003.
13. T. Preusser and M. Rumpf. Anisotropic nonlinear diffusion in flow visualization. In *IEEE Visualization 1999*, pages 325–332, San Francisco, CA, 1999.
14. G. Scheuermann, Krüger H., M. Menzel, and Rockwood A. Visualizing Nonlinear Vector Field Topology. *IEEE Transactions on Visualization and Computer Graphics*, 4(2):109 – 116, 1998.
15. G. Scheuermann, I.J. Kenneth, and W. Kollmann. Visualizing Local Vector Field Topology. *Journal of Electronic Imaging*, 9:356–367, 2000.
16. A. Telea and J.J. van Wijk. Simplified representation of vector fields. In *IEEE Visualization 1999*, pages 35–42, San Francisco, CA, 1999.
17. H. Theisel, T. Weinkauf, H.C. Hege, and H.P. Seidel. Saddle Connectors - An Approach to Visualizing the Topological Skeleton of Complex 3d Vector Fields. In *IEEE Visualization 2003*, pages 225 – 232, 2003.

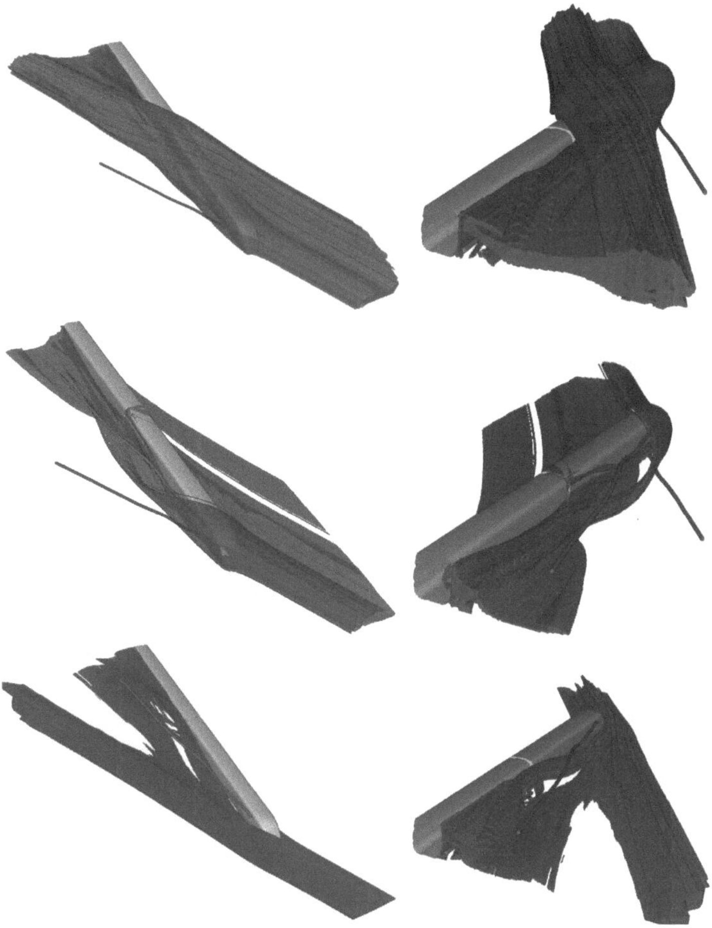

Fig. 4. Boundaries of characteristic sets of $\mathcal{G}_{Vortex}$ characterizing these parts of the flow that stream only in one vortex region (top down): $R \wedge \bar{G} \wedge \bar{B}$, $\bar{R} \wedge G \wedge \bar{B}$, $\bar{R} \wedge \bar{G} \wedge B$ (colorplate on p. 206).

18. X. Tricoche, T. Wischgoll, G. Scheuermann, and H. Hagen. Topological Tracking for the Visualization of Timedependent Two-Dimensional Flows. *Computers & Graphics*, 26(2):249 – 257, 2002.
19. T. van Walsum, F. H. Post, D. Silver, and F. J. Post. Feature Extraction and Iconic Visualization. *IEEE Transactions on Visualization and Computer Graphics*, 2(2):111 – 119, 1996.
20. G.H. Vatistas. New Model for Intense Self-Similar Vortices. *Experiments in Fluids*, 14(4):462–469, 1998.
21. T. Weinkauf, H. Theisel, H.C. Hege, and Seidel H.P. Boundary Switch Connectors for Topological Visualization of Complex 3d Vector Fields. In *Proceedings of the 6th Joint IEEE TCVG - EUROGRAPHICS Symposium on Visualization (VisSym 2004)*, pages 183 – 192, 2004.

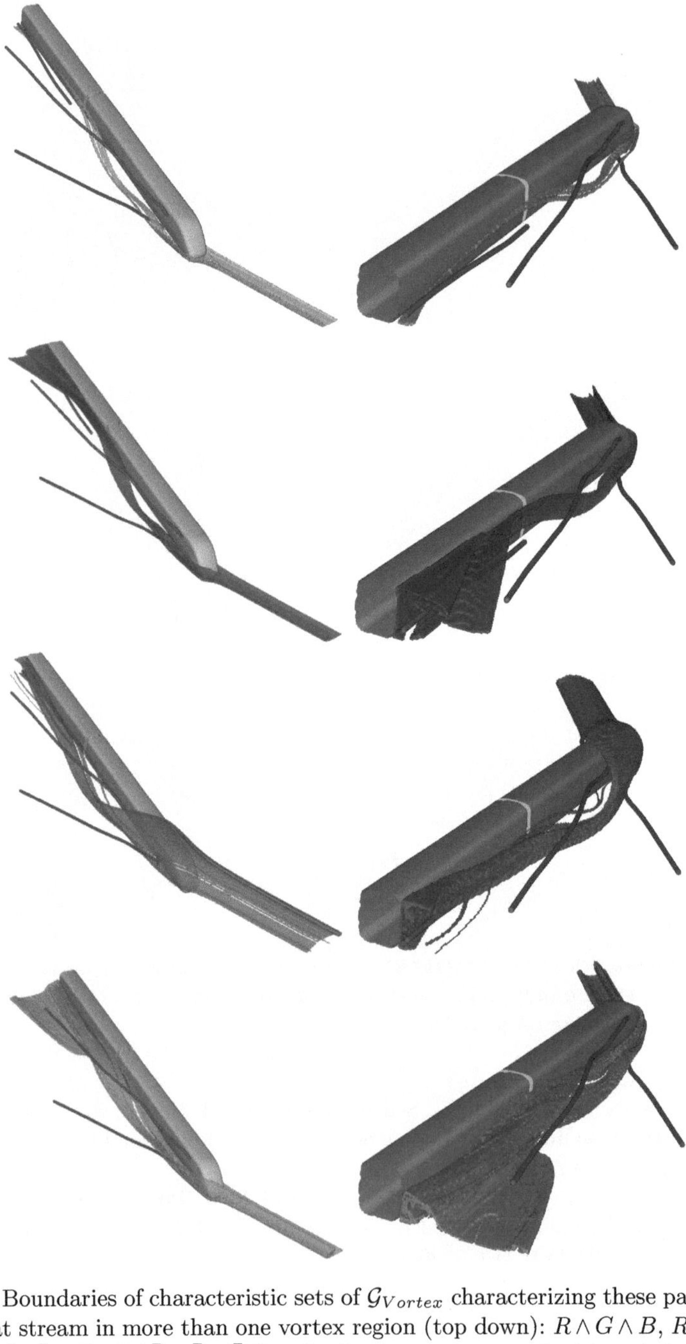

Fig. 5. Boundaries of characteristic sets of $\mathcal{G}_{Vortex}$ characterizing these parts of the flow that stream in more than one vortex region (top down): $R \wedge G \wedge B$, $R \wedge \bar{G} \wedge B$, $R \wedge G \wedge \bar{B}$, $\bar{R} \wedge G \wedge B$ ($\bar{R} \wedge \bar{G} \wedge \bar{B}$ (i.e. flowing in no vortex region) is not shown.) The second column shows the same characteristic sets from another view (colorplate on p. 207).

22. A Wiebel, C Garth, and G Scheuermann. Localized Flow Analysis of 2d and 3d Vector Fields. In Ken Brodlie, David Duke, and Ken Joy, editors, *Proceedings of Eurographics/IEEE-VGTC Symposium on Visualization 2005 (EuroVis 2005)*, pages 143–150, 2005.
23. T. Wischgoll and G. Scheuermann. Detection and Visualization of Closed Streamlines in Planar Flows. *IEEE Transactions on Visualization and Computer Graphics*, 7(2):165 – 172, 2001.

Topology-based versus Feature-based Flow Analysis – Challenges and an Application

Helwig Hauser[1], Robert S. Laramee[2], and Helmut Doleisch[1]

[1] VRVis Research Center, Vienna, Austria
{Hauser,Doleisch}@VRVis.at
[2] Department of Computer Science, Swansea University, UK
r.s.laramee@swansea.ac.uk

Summary. This paper is the result of research and contemplation on the actual usefulness of topology-based methods in real-world applications. We recapitulate commonly used arguments in favor of topology-based approaches first to realign our expectations with respect to the utilization of topology extraction in the context of concrete applications. To illustrate some of our considerations, we take a closer look at one specific example, i.e., the visual analysis of flow through a cooling jacket and we report our respective experiences. After discussing the topology-based analysis of the cooling jacket case, we contrast topology-based flow visualization with an alternative approach, i.e., the interactive feature extraction for feature-based visualization. Without generalizing just from the one concrete example scenario, we still are able to conclude with some broader experiences which we have made in the past and which seem to align well with the opinion of others in our field.

1 Introduction

Due to the rapidly increasing use of computational flow simulation and due to the concurrently increasing size and complexity of flow simulation results, there is a great demand for tools which help with the visual exploration, analysis, and presentation of flow simulation results. A vast number of technological approaches have been proposed during the last decades [9, 10, 12, 13].

Approaches range from the direct visualization of flow data, e.g., through the use of color coding or hedgehog plots, through geometric and texture-based approaches, to the large class of solutions which utilize a significant amount of computational analysis before the actual visual representation.

In this last class of flow visualization techniques, topology-based approaches are very popular. Before visualization, the topological skeleton of the flow is extracted. Critical points are identified and classified as well as critical structures of higher order (e.g., cycles, invariant tori, etc.). Their critical structures are related to each other, separatrices are computed. The extraction of the topological skeleton of flow data can be interpreted as the segmentation of the flow into regions of coherent long-term behavior – all points within a flow region which is bounded by critical structures and the associated separatrices share a common long-term behavior (at least qualitatively).

Once the topological skeleton of a flow is extracted, it can be used for visualization. In many cases [10], "only" the topological skeleton is visualized (instead of the original data). This results in a number of advantages which motivate the use of topology-based visualization techniques:

1. The extraction of a topological skeleton equals flow abstraction. Instead of the original data, *information about the flow data* is visualized. Accordingly, such a topology-based flow visualization provides information,
   - which is not explicitly contained in the original data, but abstracted from it, i.e., it is something additional to the original data,
   - which is very informative in case the long-term evolution of the flow under investigation is of significant interest, and
   - which is extremely condensed as compared to the original data (a few geometric structures convey a lot about the entire flow).

   As a result, *deeper insight* is possible through the use of topology-based flow visualization.
2. Due to the condensation as a result from the abstraction process technical advantages are yielded. Rendering a concise topological skeleton instead of millions of simulation cells allows for interactive and real-time rendering. It also reduces memory requirements significantly, i.e., the resource requirements for visualization (not necessarily for the extraction process) are drastically reduced, usually by several orders of magnitude.

These advantages motivate the use of topology-based methods for flow analysis in practical applications. If insights are possible which otherwise are impossible or especially hard to derive, qualitative benefits may result in a practical application. If interactive visualization becomes possible, even on customer computers, quantitative benefits are possible.

However, we still cannot report a wide-spread establishment of topology-based methods in real-world applications. This could be due to the fact that still methods are considered to be relatively young (many from the last decade only). However, other reasons for the fact that topology-based methods have not (yet) conquered the offices of practitioners are possible, too. In the following, we aim at a better understanding of this situation by considering a concrete example and by recapitulating additional experiences.

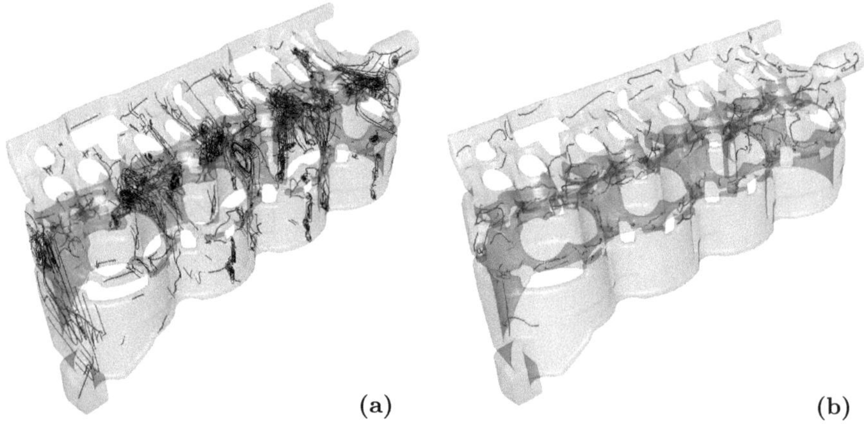

Fig. 1. (a) Cutting plane topology, revealing flow structures perpendicular to the dominant longitudinal flow – within the cut planes, streamlines are depicted in blue, longitudinal vortex cores (connected critical points) are shown in red. (b) Vortex core line extraction using the method from Sujudi and Haimes [16] (colorplate on p. 208).

2 Analyzing a Cooling Jacket, a sample from industry

To investigate the interesting question of why topology-based approaches are no yet wide-spread, we first consider a concrete example from our cooperation with AVL List GmbH in Graz, Austria, i.e., the flow through a cooling jacket as employed for the cooling of car engines [8].

Figure 1(a) shows a visualization which is based on the extraction of topological features on cutting planes through the 3D cooling jacket flow [17]. Connected critical points (in red) indicate vortical flow structures and additional streamlines in the cutting planes (depicted in blue) add in more information about the local flow structures[3]. The cutting planes are placed equidistantly along one direction chosen by the user. With some *a priori* knowledge, this approach indeed can provide useful insights into flow structures.

The application to the cooling jacket is straightforward in our case, since interesting flow structures are expected to be orthogonal to the longitudinal constituent. Positioning the cutting planes across the jacket, accordingly, reveals a number of interesting features, most notably several vortices in the head. They show up as sources and sinks on the cutting planes. The vortex cores are indicated by the connection of the critical points across the planes. This type of visualization (many line-type features in 3D) does not always yield good spatial perceptibility. The use of tube-like primitives (as one standard solution to this problem) is prohibitive in our case due to the large

[3] For original, high resolution images, please visit `http://www.VRVis.at/scivis/laramee/topology/`

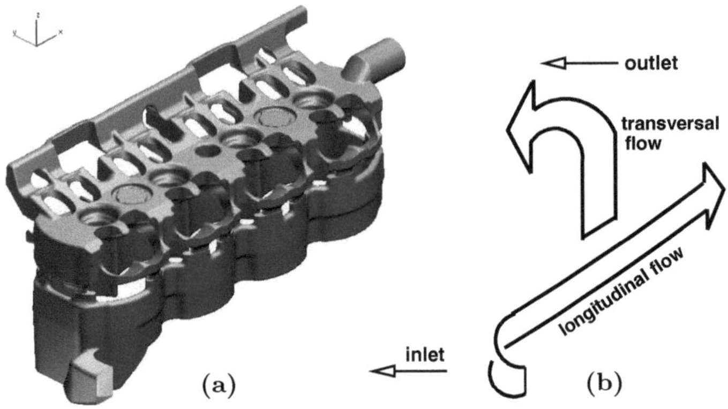

Fig. 2. The major components of the flow through the cooling jacket include a longitudinal component and a transversal component in the upward-and-over direction (from inlet to outlet).

number of lines both from a rendering performance point of view and because of visual clutter. We have therefore employed a simple scheme to illuminate the lines based on tangents, similar to the illuminated streamlines technique [15]. Nevertheless, this visualization still suffers from a lot of visual clutter and increased structural complexity so that gaining new and deep insight into the shown flow remains a challenge, even for experienced practitioners.

When comparing this topology-based visualization with a feature-based visualization, e.g., the vortex core line extraction according to Sujudi and Haimes [16] (see Fig. 1(b)), we can see that the feature-based approach generates less clutter (in this particular example), but still suffers from similar perceptual problems. However, we also know from other applications, that topology-based methods, when applied to real-world CFD simulation data, often generate a lot of geometrical structure which seems to be more difficult to control then results from feature-based approaches.

One way to address this complexity issue is to apply a topological simplification algorithm after the extraction stage. This reduces the number of critical points. Simplification algorithms of this kind often are fairly complicated to implement. Additionally, the advantage of reducing the geometric complexity of the visualization result comes at the price of an increased interpretation load on the user side – often it is not truly intuitive to understand what the simplification algorithm did (what structures have been removed and why).

3 An Engineer's Point of View

The above mentioned challenges (complexity of implementation, challenging interpretation) motivate an alternative approach to flow analysis, i.e., a more semantics-based, interactive, feature-extraction approach, guided by the engineer. It's been our experience that engineers analyzing CFD simulation data do not necessarily think in terms of critical points within the flow or in terms of a topological skeleton, but rather in terms of an ideal or optimal pattern of flow they are trying to achieve.

As an example, Fig. 2(b) attempts to depict the ideal pattern of flow through the cooling jacket geometry. A diagram similar to this one was shown to us by a mechanical engineer when we were learning what flow the CFD engineers were trying to create. There are two major components to the flow: longitudinal and transversal. Longitudinal flow is oriented lengthwise along the cooling jacket geometry whereas transversal flow is oriented in the upward-and-over direction. Essentially, the ideal pattern of flow is the most efficient path from inlet to outlet. By following the ideal path, the cooling jacket is most effective in its job of transferring heat away from the engine block. Some questions that the engineers designing the cooling jacket are interested in answering are:

1. Are there any areas where the flow is moving in the wrong direction?
2. Where, in the cooling jacket, are the areas of stagnant flow?

The function of a cooling jacket is to transfer heat away from the engine as efficiently as possible. Engineers are interested in learning where and how the flow deviates from the ideal. Deviance from the ideal leads to less effective heat transfer.

4 Interactive, Feature-based Flow Visualization

We have applied both an automatic topology extraction algorithm as well as an interactive, feature-based approach in order to investigate and analyze the behavior of the flow through a cooling jacket. The interactive, feature-based flow analysis system we have used is called SimVis [2, 5, 6]. SimVis establishes an interactive visual analysis loop with the following essential properties:

- The user visualizes the multi-variate attribute space of a CFD dataset in such a way that it can be intuitively explored and accessed directly by brushing (brushing means that interesting flow features are directly marked up in the views).
- A sophisticated interaction framework is provided that allows the user to identify interesting flow features intuitively and easily, even if the features can only be characterized in a complex specification with multiple flow attributes [3]. This includes an iterative refinement process. Specifications can be rapidly prototyped with immediate feedback.
- Linking attribute visualization to spatio-temporal visualization yields feature-based focus+context visualization of the CFD dataset in an intuitive manner.

(a) (b)

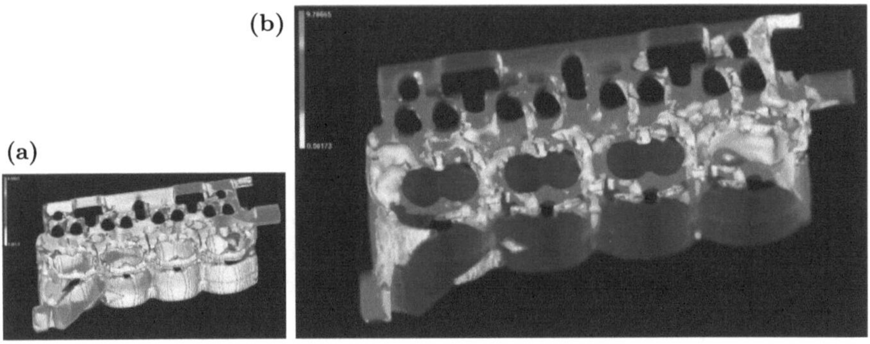

Fig. 3. (**a**) The visualization of all regions of forward-longitudinal flow; color-mapping reflects velocity magnitude. (**b**) The result of selecting all regions of reverse-longitudinal flow, the inverse of the left selection (colorplate on p. 208).

(a) (b)

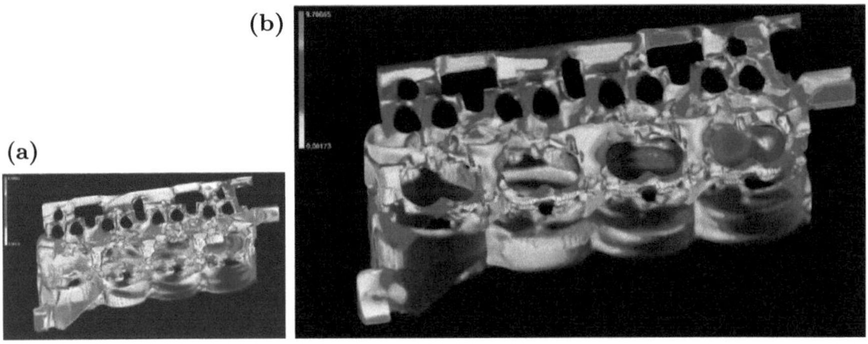

Fig. 4. (**a**) The visualization of all regions of forward-transversal flow; color-mapping indicates velocity magnitude. (**b**) The result of selecting all regions of reverse-transversal flow, again the inverse of the left selection (colorplate on p. 208).

- A feature specification system is provided that (in contrast to most other related approaches) reflects the often quite smooth distribution of flow attributes across the domain, a property resulting from most CFD flow simulations. This is achieved via degree of interest functionality instead of using sharp selections [4].

SimVis uses multiple linked views, utilizing visualization techniques of different kind (scatterplots, histograms, 3D visualization, etc.). Generally, the user specifies which subsets of the flow to focus on, e.g., by brushing in a scatterplot. The marked data subsets are then rendered as the visually emphasized focus in a focus+context visualization style. In the following, we describe the use of SimVis in order to answer the questions outlined above by the engineer.

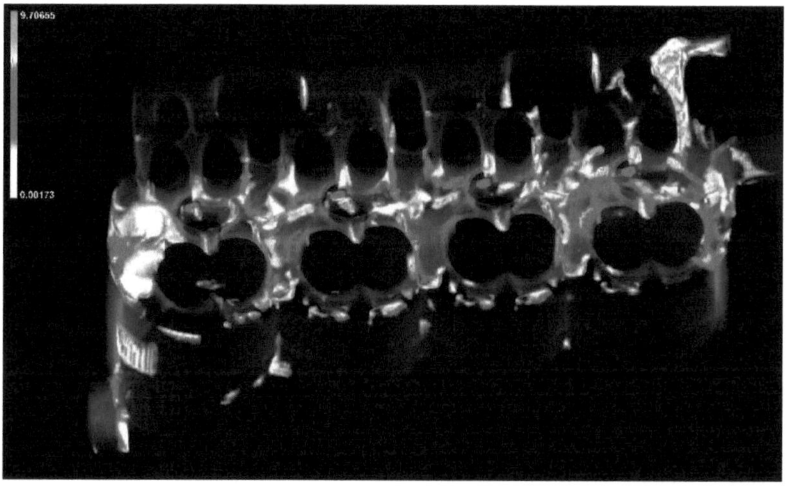

Fig. 5. The result of selecting all regions of reverse-longitudinal flow *and* regions of reverse-transversal flow (colorplate on p. 209).

Extracting Forward and Reverse-Longitudinal Flow

Figure 3(a) depicts the result of selecting all positive x-velocity values via smooth brushing. The positive x-axis is aligned with the longitudinal flow direction. Thus all regions containing a positive x-velocity are flowing, at least partially, forward with the goal of traversing the shortest path from inlet to outlet. Figure 3(b) shows the result of selecting all negative x-velocity values. The right image is more interesting to the engineer. It shows precisely those regions in the geometry where the flow is moving in the opposite direction from that which is desired. Here, the forward-longitudinal flow fills 76.2% of the volume ($\pm 1\%$) whereas reverse-longitudinal flow occupies 18.4% ($\pm 1\%$). This is not a bad forward-to-reverse flow ratio, but still significantly different from the ideal. It is interesting, for example, to see a region of reverse-longitudinal flow immediately near the inlet, representing a major recirculation zone.

Extracting Forward and Reverse-Transversal Flow

Figure 4 visualizes both forward and reverse regions of transversal flow. Essentially, engineers are mostly interested in seeing where in the geometry the ideal pattern of flow is not being realized, e.g., in the right image. Figure 4(b) reveals many regions of flow that are traveling downward, or rather against the ideal current of flow. In fact, the amount of forward-transversal flow is only 54.6% ($\pm 1\%$) while the amount of reverse-transversal flow is 37.7% ($\pm 1\%$). The amount of reverse-transversal flow is considerably higher than what we expected. A major region of reverse-transversal flow, for example, can be seen in the second cylinder block (from the left). We see regions of reverse-transversal flow again at the inlet.

Fig. 6. A feature-based, focus+context visualization showing regions of near-stagnant flow, specified interactively (colorplate on p. 209).

We can further refine the region of interest by including only velocity values with negative x- *and* negative z-components. Figure 5 depicts the regions where flow moves backward and down instead of the shortest path – up and forward from inlet to outlet. From this result, we can deduce that flow through the cylinder head is a complex patchwork of flow, especially along the center of the head.

Extracting Regions of Stagnant Flow

Figure 6 illustrates regions with a velocity value, $|\mathbf{v}|$, of less than 0.1 m/s. We know that regions of stagnant flow, like those in Fig. 6, are less effective in transporting heat away from the engine. The color-coding in Fig. 6 indicates temperature. The optimal fluid temperature, $363°K$, is mapped to green and higher temperatures are mapped to red. This visualization result indicates that there are very few, small regions where low velocity and high temperature coincide – an advantageous design characteristic for an engine part designed to transport heat away. Figure 7 further refines the feature specification by also restricting the focus to high temperature values. The new feature is defined as:

$$(|\mathbf{v}| < 0.1\, m/s) \cap (364°K < t)$$

The result in Fig. 7 is a less cluttered image, showing undesirable regions, where slow flow and hot flow are apparent. These regions are less effective in transporting heat away. Fortunately, these regions seem to be rather small, thus, from a heat-transfer point of view, the simulation results point toward a good design. Areas of very high velocity, leading to cavitation, can be identified in a similar way.

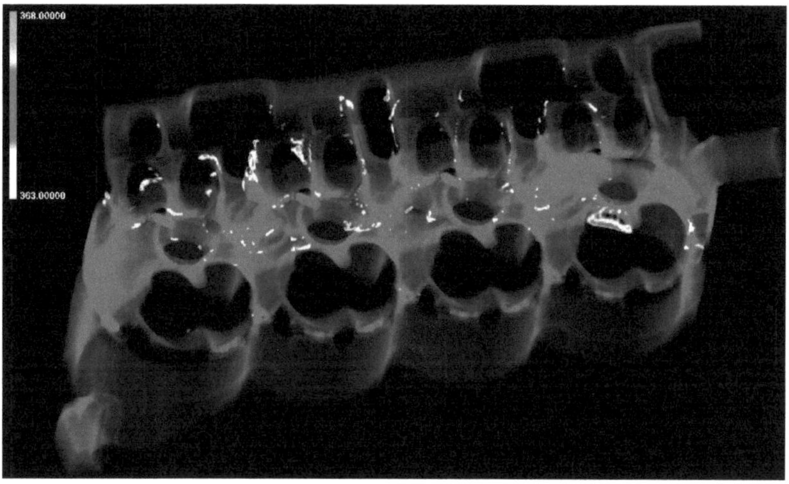

Fig. 7. Areas of temperature $t > 364\,^{\circ}K$ and velocity $|\mathbf{v}| < 0.1\,m/s$ are interactively-specified by the user and rendered in focus (colorplate on p. 209).

Considering the above described analysis, we state, that obviously also other options for the investigation of the properties of this particular application scenario exist. Alternatively, an engineer also could pick a simulated measure of heat-flux and analyze the data with respect to this data attribute. The Nusselt number [19] or the heat transfer coefficient [1] could be used if available from the simulation.

5 Semantics-based Segmentation of Flow

What we have done in Sect. 4 is essentially a segmentation of the flow based on semantics. Such semantics-based segmentation is standard practice in many other fields, e.g., in medical visualization. A notion of objects or regions (or the like) is generated on top of the raw data which is more meaningful to application experts.

In the above discussed example, the flow domain has been segmented according to different components of the flow direction. In this particular context, each of the Cartesian velocity direction has a specific meaning. The semantics are based on the questions posed by the engineer in Sect. 3. One speciality here, however, as compared to other standard domain segmentation approaches, is that smooth region boundaries are considered (smooth brushing [4]). Another speciality here is that we have employed multi-variate segmentation in our analysis, i.e., selections with respect to multiple attributes.

When we compare this approach with the classical topology-based approaches, we claim that interpretation of the results is easier and clearer to the user (at least in many cases). Improved comprehension of the visualization seems to result from (a) a step-by-step extraction process with steps easy to understand and (b) an approach which per se associates well with the way the users think about their data (in terms of the attributes rather than in complex terms of flow topology).

6 Topology-based Flow Visualization in Industry

Unlike direct flow visualization such as color-coding of velocities, or geometric techniques like the use of streamlines or isosurfaces, topology-based flow visualization methods not really have made their way into common commercial flow visualization software. Why not? In a manner inspired by Globus and Raible [7], we try to list some possible answers to this question. We note that a similar theme is addressed by Van Wijk [18] about the value of scientific visualization in a broader sense.

1. The advantage of extracting meaningful, high-level abstractions from flow data (such as topological features) at the same time seems to be a disadvantage, also: Higher levels of abstraction are more difficult to understand and can cause problems with their interpretation. More cognitive work is required at the user side.
2. Topology-based methods usually are not easy to implement. The extraction of topological features can be challenging in unstructured grids, in higher dimensions, from noisy datasets, and also with respect to robust numerics. From a computational point of view, for example, it is very challenging to properly compute the separatrix structure of a 3D vector field [11]. When Galilean-invariant solutions are required [14], many existing techniques fail. Also, the dependence of extraction results on the turbulence models employed in the simulation makes interpretation difficult in some cases.
3. Development and use of topology-based methods are costly, including:
   - An initial development cost, including one or more engineers, possibly also the acquisition of new hardware.
   - An initial cost per user – topology-based analysis techniques usually are not intuitive to use; also, in a highly specialized environment, tools sometimes lack optimal GUIs and their usage requires special training.
   - Costs per session/use, including the time it takes the user to generate a visualization from a given algorithm or method each time of use.
   - The cost of cognition, i.e., the time the user needs to understand and explore the visualization result to gain insight into the underlying data.

   Taking into account that many application questions also are solvable with more simple approaches, costs also might be responsible, why topology-based methods are not so often used in practive.
4. This tightly relates to the question whether a topology-based solution is a “must have” or a “nice to have” – it is much harder for the latter to

establish itself on the market; in this case the cost-question (see above) plays a much more important role than in the "must have" case.

5. As already mentioned, it is often also possible to solve the same problem with other, more simple methods. Given a suitable seeding strategy, streamlines may be used to visualize critical points in a planar domain, for example. An explicit extraction computation is not always required.
6. Lack of communication between communities also might be responsible. These days, there is quite a gap between the visualization research community and prospective users (they usually do not visit the visualization conferences, for example). In fact, other communities such as the engineering analysis community are not even aware that a visualization community *exists*. Closely related is the lack of inter-community knowledge transfer and a lack of educational literature.
7. Lack of customer demands: Software development in industry usually is driven by customer demands, i.e., customers who demand new features of the software.

Surely, some of these problems can be solved. The last problem on the list, for example, might be addressed when more motivation for inter-disciplinary communication is generated. The problem of difficult implementation may be solved with more time invested in research and development. Often algorithms, which present an easier and more elegant solution to a problem that originally required a very complex solution, are published at later points in time in a larger community. Clearly, visualization solutions with less complexity are needed as well as more communication between fields of expertise.

The above list summarizes some of the challenges that topological methods must face before being incorporated on a more wide-spread basis. Does this mean we should stop topology-based visualization research? No. The original motivations for this line of work are more prevalent than ever. Data sets grow at faster rates than hardware and this trend promises to continue. Data, especially CFD simulation data, is becoming ever more complex. The demand for tools that can help the user sift through this complexity will only increase. The research field of topological analysis is still relatively young, thus much work remains to be done.

Acknowledgments

The authors thank all who have contributed to this work, including AVL (`www.avl.com`) and the Austrian research program K*plus*. We thank Jürgen Schneider (AVL) for collaborating as application partner. We thank Christoph Garth for helping with the implementation. We thank Stephan Schmidt and Jarke J. van Wijk for their thoughts on why topology-based methods are not so much used in industry. We thank Frits H. Post and Benjamin Vrolijk for their input with respect to feature-based flow visualization. We thank Philipp Muigg for his help with the SimVis system (`www.SimVis.at`). The CFD simulation data is courtesy of AVL.

References

1. A. Bejan and A. Kraus. *Heat Transfer Handbook*. John Wiley & Sons, 2003.
2. H. Doleisch. *Visual Analysis of Complex Simulation Data using Multiple Heterogeneous Views*. PhD thesis, Vienna University of Technology, Austria, 2004.
3. H. Doleisch, M. Gasser, and H. Hauser. Interactive Feature Specification for Focus+Context Visualization of Complex Simulation Data. In *Data Visualization, Proc. 5th Joint EG – IEEE TCVG Symp. on Vis.*, pages 239–248, 2003.
4. H. Doleisch and H. Hauser. Smooth Brushing for Focus and Context Visualization of Simulation Data in 3D. In *Proc. WSCG 2002*, pages 147–151, 2002.
5. H. Doleisch, M. Mayer, M. Gasser, P. Priesching, and H. Hauser. Interactive Feature Specification for Simulation Data on Time-Varying Grids. In *Proceedings of Simulation and Visualization 2005 (SimVis 2005)*, pages 291–304, 2005.
6. H. Doleisch, M. Mayer, M. Gasser, R. Wanker, and H. Hauser. Case Study: Visual Analysis of Complex, Time-Dependent Simulation Results of a Diesel Exhaust System. In *Data Visualization, Proc. VisSym*, pages 91–96, 2004.
7. A. Globus and E. Raible. Fourteen Ways to Say Nothing with Scientific Visualization. In *IEEE Computer*, pages 86–88, 1994.
8. R. S. Laramee, Chr. Garth, H. Doleisch, J. Schneider, H. Hauser, and H. Hagen. Visual Analysis and Exploration of Fluid Flow in a Cooling Jacket. In *Proceedings of IEEE Visualization 2005*, pages 623–630. IEEE Computer Society, 2005.
9. R. S. Laramee, H. Hauser, H. Doleisch, Fr. H. Post, B. Vrolijk, and D. Weiskopf. The State of the Art in Flow Visualization: Dense and Texture-Based Techniques. *Computer Graphics Forum*, 23(2):203–221, 2004.
10. R. S. Laramee, H. Hauser, H. Doleisch, Fr. H. Post, and L. Zhao. Topology-based Flow Visualization, the State of the Art. In *Topology-based Methods in Visualization*. Springer. (chapter 1 of this book).
11. K. M. Mahrous, J. C. Bennett, B. Hamann, and K. I. Joy. Improving Topological Segmentation of Three-dimensional Vector Fields. In *Data Visualization, Proc. 5th Joint EG – IEEE TCVG Symp. on Vis.*, pages 203–212, 2003.
12. Fr. H. Post, B. Vrolijk, H. Hauser, R. S. Laramee, and H. Doleisch. Feature Extraction and Visualization of Flow Fields. In *Eurographics 2002 State-of-the-Art Reports*, pages 69–100, 2002.
13. Fr. H. Post, B. Vrolijk, H. Hauser, R. S. Laramee, and H. Doleisch. The State of the Art in Flow Visualization: Feature Extraction and Tracking. *Computer Graphics Forum*, 22(4):775–792, 2003.
14. J. Sahner, T. Weinkauf, and H.-Chr. Hege. Galilean Invariant Extraction and Iconic Representation of Vortex Core Lines. In *Proceedings of the Joint Eurographics - IEEE VGTC Symposium on Visualization (EuroVis 05)*, 2005.
15. D. Stalling, M. Zöckler, and H.-Chr. Hege. Fast Display of Illuminated Field Lines. *IEEE Transactions on Visualization and Computer Graphics*, 3(2), 1997.
16. D. Sujudi and R. Haimes. Identification of Swirling Flow in 3D Vector Fields. Technical Report AIAA Paper 95–1715, Am. Inst. of Aeron. & Astron., 1995.
17. X. Tricoche, Chr. Garth, G. Kindlmann, E. Deines, G. Scheuermann, M. Rütten, and H. Hagen. Visualization of Intricate Flow Structures for Vortex Breakdown Analysis. In *Proc. IEEE Visualization 2004*, pages 187–194, 2004.
18. J. J. van Wijk. The Value of Visualization. In *Proceedings of IEEE Visualization 2005*, pages 79–86. IEEE Computer Society, 2005.
19. F. M. White. *Heat Transfer*. Addison-Wesley, 1984.

Topology Based Flow Analysis and Superposition Effects

Julia Ebling[1], Alexander Wiebel[1], Christoph Garth[2], and Gerik Scheuermann[1]

[1] Universität Leipzig
{ebling,wiebel,scheuermann}@informatik.uni-leipzig.de
[2] Technische Universitt Kaiserslautern
garth@informatik.uni-kl.de

Summary. Using topology for feature analysis in flow fields faces several problems. First of all, not all features can be detected using topology based methods. Second, while in flow feature analysis the user is interested in a quantification of feature parameters like position, size, shape, radial velocity and other parameters of feature models, many of these parameters can not be determined using topology based methods alone. Additionally, in some applications it is advantageous to regard the vector field as a superposition of several, possibly simple, features. As topology based methods are quite sensitive to superposition effects, their precision and usability is limited in these cases. In this paper, topology based analysis and visualization of flow fields is estimated and compared to other feature based approaches demonstrating these problems.

1 Introduction

Visualization and analysis of vector fields from flow simulations and measurements is an important step in engineering processes, e.g. during the design phase of airplanes, cars, trains, and combustion chambers. According to Dr. Shneiderman's Visual Information Seeking Mantra [13] visualization of data consists of three steps: "Overview first, zoom and filter, then details-on-demand". However, due to the extreme size and complexity of today's data sets it may not be possible to provide an overview of the data. Interactive zooming and filtering also quickly reaches its limits. Thus the pipeline has to be extended by an automatic analysis step to reduce size and/or complexity of the data before trying to visualize it. This results in Keim's Visual Analytics Mantra "Analyse first, show the important, zoom, filter and analyse further, details on demand" that he proposed at the Workshop on Visual Analytics 2005 in Darmstadt.

Up to now topology has mainly been used to give an overview of a flow as it segments a data set into regions of same flow behavior. In this paper,

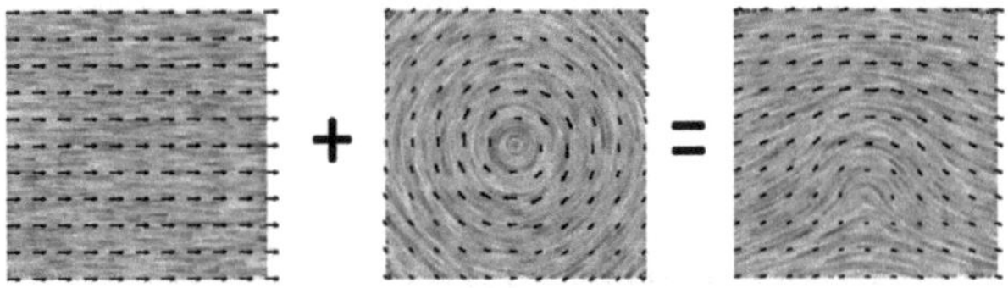

Fig. 1. Linearity of features - the superposition principle: The flow on the right is the direct sum of the two different flows on the left.

advantages as well as deficits of topological methods concerning the analysis step mentioned in the previous paragraph are discussed. Therefore, topology is compared to feature based methods. We do not only treat topology as a description of the flow, but also analyze topology within the context of typical feature models like the Vatistas vortex [17]. Concerning this model the ability of topology based methods to determine parameters of feature models is analyzed (Section 3).

For signal processing, linear, shift invariant signals play an important role as most signals can be described, or at least approximated, by them. Other signals can be approximated quite well, too. The linearity property is also known as the superposition principle, stating that a combination of signals is equal to the same combination applied to all parts of these signals. This means that a complex signal can be understood as a linear combination of several simpler signals (Figure 1). Keeping this signal processing perspective in mind, we can treat vector fields as superpositions too and thus can describe phenomena like vortices hidden in a strong homogeneous flow. As topology is based on critical points, that is points where the velocity in the field is zero, it is quite sensitive to changes of the mean flow of a data set. Adding or subtracting, i.e. superposing, different constant flows to a flow, thus will change its topology. One way to avoid these changes is to analyze the localized or region-specific flow [18] which is independent of constant and homogenous flows passing through the considered region. Superposition effects, however, can also appear independent of such flow components, e.g. when two vortices overlap. In Section 4 we discuss superposition phenomena and their influence on topology based methods and compare the superposition view to the usual perception of vector fields describing the actual flow.

2 Feature Definitions

Features are often defined as "phenomena, structures or objects in a data set, that are of interest for a certain research or engineering problem" [9]. It is not possible to give a list of all features of interest for flow fields in general as these differ from application to application and small changes of one feature can lead to a variety of new features. Nevertheless, most features can be categorized into a few groups like vortices and other swirling flows, shock waves, shear flow and boundary layers, reversed flow, saddle points, separation and attachment lines or surfaces, areas with convergent or divergent behavior, and regions with homogeneous or constant flow. The remainder of this section gives a short overview of feature definitions we will use in the following discussion.

2.1 Vector Field Topology

In a vector field, as already mentioned, positions where the magnitude of the vectors is zero are called critical points. Critical points can be classified by the eigenvalues and eigenvectors of the velocity gradient tensor at their position. In linear 2D vector fields, they are classified as sinks, sources, saddles and center points. Sinks and sources can be of the types focus, spiral, node or improper node. Critical points are connected by streamlines called separatrices which divide the vector field into regions of same flow behavior. The topology graph of a vector field consists of all critical points, separatrices and closed streamlines of the vector field.

2.2 Correlation Methods

Vortices, shear flow, sources, sinks, saddles, separation and attachment lines, and homogeneous flow can all be coded into vector valued masks (for examples see Figure 2). Using these masks for (rotation invariant) pattern matching based on correlation allows to detect similar areas in a vector field [2, 5]. A disadvantage of this approach is the computational time for irregular grids. For regular grids, acceleration via fast Fourier transform can be used [3]. The main advantage of this approach is the robustness to noise due to the averaging in the correlation. This is of great importance for measured data. Furthermore, as smoothing can be described by a convolution with a scalar valued mask, it is commutative with the pattern matching. When the average of a mask is zero, the matching results are independent of any mean flow of the vector field. As the masks are linear, shift invariant filters they are suitable for the analysis of superposition phenomena. Feature definitions using vorticity can be integrated into this approach as derivation and derived quantities can be computed via convolution with a mask [2, 3].

2.3 Threshold Based Feature Definitions

Thresholds for high vorticity (curl of velocity, $\nabla \times v$), high helicity (projection of vorticity onto velocity, $(\nabla \times v) \cdot v$) and low modified pressure (negative λ_2, see [6]) are often used to determine position, size and shape of vortices. In addition to these, there is a variety of vortex core detection algorithms of all kinds. An overview can be found in [9, 11]. Vorticity and λ_2 are Galilean invariant, that is, they are independent of the frame of reference and thus independent of any mean flow. Computing vorticity is also commutative with smoothing, that is, computing vorticity from smoothed data yields the same results as computing it first and performing smoothing afterwards.

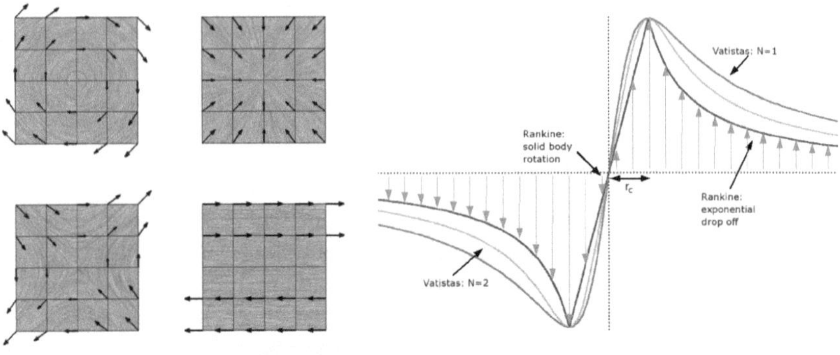

Fig. 2. Left: Rotation, convergence, saddle and shear, each visualized using hedgehogs and LIC. Right: The Vatistas vortex model. A pure, circular rotation is assumed. A cut through this rotation results in the described velocity profile (coloplate on p. 210).

3 Quantification of Feature Parameters

For detailed flow visualization, a preceding analysis of the flow and its features is often necessary. In this section, the advantages and limits of topological methods for an analysis of features are discussed and compared to those of other feature based approaches. All properties are demonstrated using the Vatistas vortex [17] (Figure 2), a popular vortex model in fluid dynamics, as an example.

3.1 Feature Models - an Example

The definition of a Vatistas vortex [17] is based on a pure, circular rotation with the magnitude of the vectors given by

$$v(r) = 2^{1/N} r_c^2 \frac{v_c r}{(r_c^{2N} + r^{2N})^{1/N}},$$

where r_c is the vortex core radius, r the distance to the vortex core, v_c the velocity at the vortex core radius, and N a parameter describing the transition from the linear velocity profile within the vortex core to the exponential drop-off of the velocity magnitudes outside. For $N = 1$, the Vatistas model equals the Scully vortex model [12] and for $N = \infty$, the Rankine model [10] is obtained. Important parameters which engineers want to determine in this model are vortex core center or vortex position, vortex core radius, the circumferential velocity at the vortex core radius, the circumferential velocity distribution, overall vorticity within the vortex core and, in 3D, the maximal axial velocity and the axial velocity distribution [1, 16].

For the Vatistas vortex, the vectors describing the flow are more important than the actual streamlines. The velocity of the flow is not only indispensable for the engineers whereas it is neglected in topological methods, but the projection of the vectors describing the actual flow onto the vectors describing the Vatistas vortex can be more important than the flow itself. Due to this projection, the topology of this vortex can differ greatly according to the remaining part of the flow (Section 4).

The Vatistas vortex is an abstraction of vortices. Though tunable by some parameters like vortex core radius and velocity distribution, this model assumes a perfectly circular vortex and thus only approximates real flow (Figure 2). The vortex is assumed to spread out infinitely, though the influence of the vortex will converge to zero with increasing distance to the center. This means that the region of significant influence will be larger than the actual vortex core and spread out over regions separated by topology. This demonstrates that the way of thinking of engineers simulating and analyzing flow is often not topology based, which may lead to problems applying topology to answer their questions concerning the flow.

3.2 Topology and Velocity

In this subsection, the relation of the velocity magnitude of the flow to the resulting topological information is analyzed. For this, we compare two streamlines in two fields with different velocity magnitudes but the same velocity directions. Let $D \subset \mathbb{R}^d$ be an open domain and $v : D \to \mathbb{R}^d$ a vector field satisfying the Lipschitz condition. Let $Z := \{ z \in D \mid v(z) = 0 \}$ be the set of critical points and $D_\varepsilon := \{ x \in D \mid \forall z \in Z : |x - z| > \varepsilon \}$ the domain without the critical points and their ε-neighborhood. In the following, we compare the two vector fields

$$\begin{aligned} \bar{v} : D_\varepsilon \to \mathbb{R}^d, \ x &\mapsto v(x) \\ \tilde{v} : D_\varepsilon \to \mathbb{R}^d, \ x &\mapsto \frac{v(x)}{|v(x)|}. \end{aligned}$$

For $a \in D_\varepsilon$ and $\bar{I} = (0, t_{max})$ let $\bar{c}_a : \bar{I} \to D_\varepsilon$ be the well-defined streamline of $\bar{v}$ through a with maximal length, that is

$$\begin{aligned} \bar{c}_a(0) &= a \\ \frac{\partial \bar{c}_a}{\partial t}(t) &= \bar{v}(\bar{c}_a(t)) = v(\bar{c}_a(t)). \end{aligned}$$

Define the following mapping for reparametrization:

$$l : \bar{I} \to l(\bar{I}), \quad l(t) = \int_0^t |\bar{v}(\bar{c}_a(\tau))| \, d\tau.$$

This function is strictly monotonic and therefore invertible. We define

$$\tilde{I} = (l(0), l(t_{max})) = l(\bar{I})$$

$$\tilde{c}_a : \tilde{I} \to D_\varepsilon, \ s \mapsto \bar{c}_a(l^{-1}(s))$$

Then $\tilde{c}_a$, a reparametrization of $\bar{c}_a$, is the well-defined streamline of $\tilde{v}$ through starting point a:

$$\tilde{c}_a(0) = \tilde{c}_a(l(0)) = \bar{c}_a(0) = a$$

$$\begin{aligned}
\frac{\partial \tilde{c}_a}{\partial s}(s) &= \frac{\partial \bar{c}_a}{\partial s}(l^{-1}(s)) \\
&= \bar{v}(\bar{c}_a(l^{-1}(s)))\frac{1}{\frac{\partial l(t)}{\partial t}} \\
&= \bar{v}(\bar{c}_a(l^{-1}(s)))\frac{1}{|\bar{v}(\bar{c}_a(l^{-1}(s)))|} \\
&= \bar{v}(\tilde{c}_a(s))\frac{1}{|\bar{v}(\tilde{c}_a(s))|} \\
&= \tilde{v}(\tilde{c}_a(s)).
\end{aligned}$$

This means that the streamlines of $\bar{v}$ are the same as the streamlines of $\tilde{v}$. Thus, the velocity magnitude of the whole field can be artificially set to one and, in contrast to vorticity and other quantities, streamlines and flow topology will stay the same. This reveals that flow topology is independent of the velocity magnitudes of a vector field and thus shows that topology is ill suited to describe or analyze features whose models strongly depend on the velocity magnitude of the flow.

3.3 Determining Parameters

The center of a pure Vatistas vortex is a critical point in 2D, and thus the position of the vortex can be easily determined using vector field topology. In contrast to other feature definitions based on vorticity or pattern matching, the center position is automatically determined with subpixel accuracy. This is a distinct advantage of topology as subpixel accuracy is often hard to obtain [4]. However, topology is sensitive to noise, therefore subpixel results are meaningless for noisy data such as those obtained by measurements where the only solution is smoothing the data.

The determination of the size of a vortex – or its vortex core region – is much more challenging. For a pure 2D Vatistas Model, no size can be determined using topological methods. There are closed orbits at all distances to the center and so separatrices in this case. When two vortices interact, often a saddle point confines the regions of the two vortices. The separatrices as defined by the saddle point seem to enclose the vortex regions. However, in the case of a spiraling separatrix, no size can be determined numerically.

Comparing these quantities with the Vatistas parameters again reveals the two different approaches taken by topology and feature based descriptions.

Topology segments the vector field into regions of same flow behavior, that is every particle within the vortex region as defined by topological methods will pass into the critical point or has emerged there. The vortex core radius as defined by the Vatistas vortex can be either larger or smaller than that region, as the topology depends entirely on the streamlines while for the Vatistas vortex, only the projection of the vectors onto a perfect rotation is evaluated (Figure 2). This aspect is detailed further in Section 4.

From this discussion it can be seen quite clearly that topology and feature models from engineering take quite different perspectives on the definition of interesting features. The first globally describes regions of same flow behavior in relation to inflow and outflow regions while the other is more interested in local properties like velocity and vorticity, which can not be determined using topology based methods at all (Section 3.2). However, it is application dependent which of these two perspectives is more appropriate or beneficial.

4 Superposition Effects

In this section, we will discuss the two different perceptions of vector fields mentioned in the previous section: the usual point of view based on streamline behavior and the perception of complex vector fields as a superposition, or linear combination, of several, possibly simpler vector fields (see Figure 1). We will call the first interpretation *interaction view* and the latter *superposition view*. Furthermore, the resulting effects for analysis and visualization using topology or other feature based methods are shown.

The difference between the two views becomes clear by looking at the occurrence of two or more features in close vicinity. The superposition view emphasizes the original features and their parameters. The interaction, on the other hand, describes how these features influence each other, by overlapping, friction and other phenomena, resulting in the actually observed flow. Up to now, nearly always only the interaction view has been studied and visualized. But the user is not only interested in the actual flow, but in explanations and, possibly simplified, models of the underlying phenomena.

In 2001 an international cooperative research program called HART II was conducted to investigate the physics of blade pressure, noise radiation, and vibrations caused by the wake of helicopter rotors [1, 4, 16]. Three-component Particle Image Velocimetry (3-C PIV) was part of the measurements. In the resulting vector fields superposition phenomena are ubiquitous (Figure 3) due to the overall velocity of the flow in the wind tunnel. Furthermore, each crossing of a blade creates, among other things, a new vortex which is added to the flow created by previous blade crossings, the movement and the shape of the helicopter. To understand the wake of the rotor blades, and to be able to create a model of it, all vortices and other features have to be detected and their parameters have to be determined. For accurate determination of the

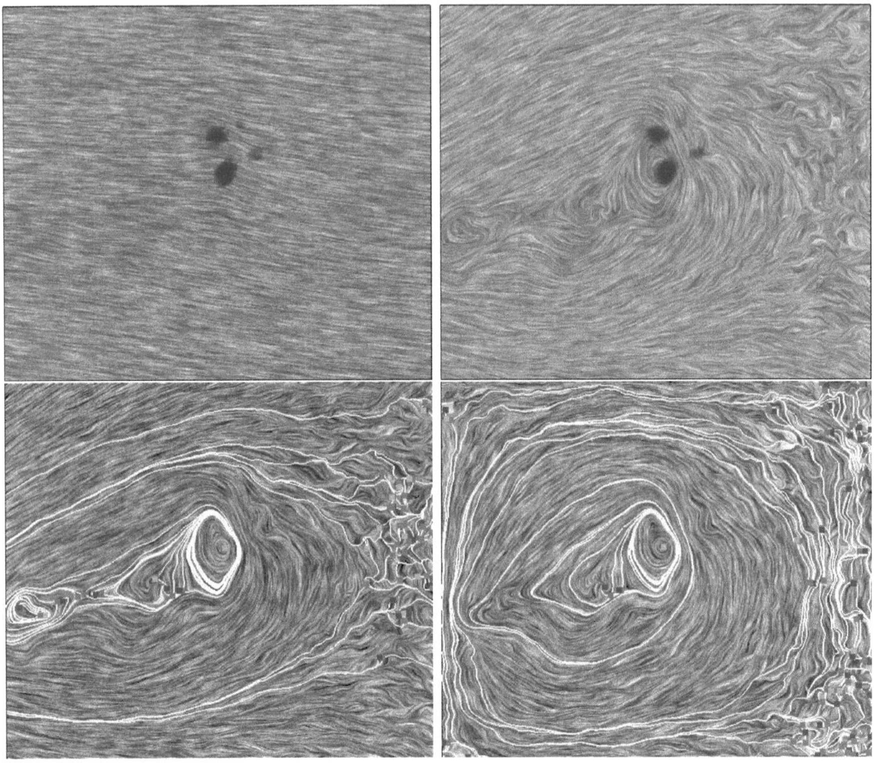

Fig. 3. One vector field of the HART II test measurements. Top left: LIC and vorticity of the original data set, dark blue: high negative vorticity, red: high positive vorticity. Top right: Vorticity and LIC of the data set after removing the average. Bottom left: LIC and topology after removing the average. Bottom right: LIC and topology of the region-specific flow of the dataset (colorplate on p. 210).

parameters, the superposition effects and their consequences for the accuracy of the analysis methods have to be studied.

4.1 Localized Flow Analysis

Superposition is an effect most common in wind tunnel measurements. Dominant passing flow induced by the blower often hides vortices and other features so a direct visualization may not reveal all features or even none at all [4, 15]. Regarding vortex detection this is most often dealt with by using vorticity or λ_2 as these quantities are Galilean invariant and thus do not vary with added or subtracted constant flow. Another method is to compute the average and remove it from the vector field. But the vortices, though having zero average, can add to the average of the whole field as they are assumed to be spread out infinitely and only a part of the vortex, not having zero average, might be within the data set. Thus, removing the average will change the results of a latter analysis of the data. Furthermore, different vortices may appear or disappear when subtracting or adding different constant flows (see [15] and Figure 3).

One approach to solve this problem is to divide the flow field into three fields containing the divergence, rotation and harmonic parts using the Hodge decomposition theorem [7, 8, 14]. Features are then detected as extremal points of the divergence and rotational field. However, it is not quite clear how analysis and visualization methods are effected by this decomposition.

A better solution is to remove the boundary induced flow to get a region-specific or localized flow as discussed in [18]. The flow through the boundary of the region-specific field is zero. Vorticity and divergence, and thus the local features of the original flow, are preserved in the region-specific flow and can be visualized using topology (Figures 3 and 4). The region-specific flow thus is independent of superposed constant and homogenous flows and represents a basis with non-changing topology for fields with different superposed constant or homogenous flows.

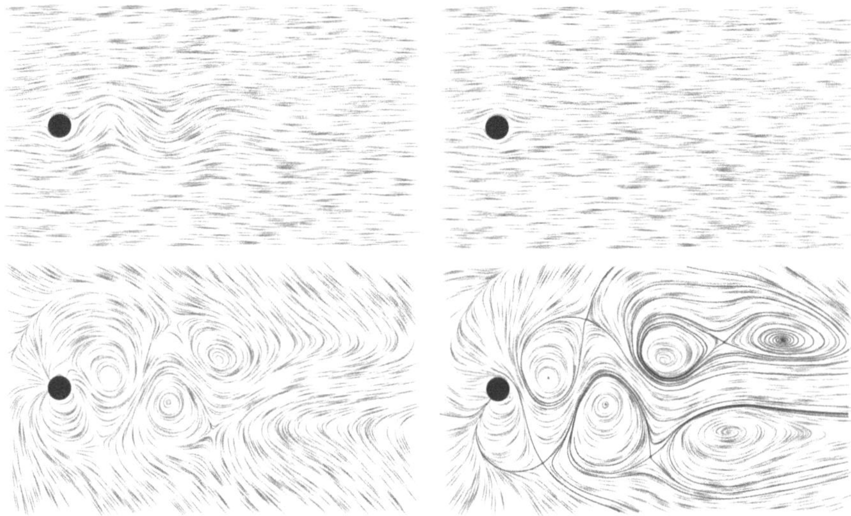

Fig. 4. Comparison of different fields obtained from a cylinder data set with a Kármán vortex street. Top left: Streamlines in the original flow. Only sinuous structures of the lines give hints on the vortices. Top right: Potential flow induced by the boundary. Bottom left: Three vortices revealed by removing the average flow. Bottom right: Subtracting the potential flow reveals all five vortices by use of topology (coloplate on p. 211).

4.2 Superposition of Nearby Features

Superposition and interaction can also take place without added constant flow, two vortices for example can influence each other strong enough to result in warped visualizations or hiding of one of the vortices. In Figure 3, the effects of removing the average or computing the localized flow are shown for one data set of the HART II test. Neither of the two techniques results in a vector field where more than the most dominant vortex is found using topological methods. However, computing the vorticity, matching with a vortex mask, or evaluating the Vatistas vortex model directly reveals the multiple vortices the engineers examine in this data [1, 16]. Therefore, local superposition effects caused the hiding of these vortices.

The authors generated some test data sets using different Vatistas vortices [17] to qualify and quantify the effects caused by superposition. The first two data sets consist of a weak vortex with radius r=2 and radial velocity $v_r = 1$ and a stronger vortex with r=5 and $v_r = 3$, both with N=1. In the first data set (Figur 5, top left), both vortices have same rotation direction, and in the second data set (Figure 5, bottom left), the rotation direction differs. In both images, stramline based approaches like topology detect one vortex only. Template matching with a 3x3 rotation mask detects both vortices.

The accuracy of current feature definitions for the determination of the parameters of the underlying features was studied as well (Figure 5, right). Here, data sets consiting of two vortices with r=5, $r_c = 1$, and N=1, were generated. Template matching with a 3x3 rotational mask detects the true vortex centers (Figure 5, top right). Setting all velocity magnitudes in the field to one, and matching afterwards, yields results more similar to the topological features (Figure 5, bottom right).

Topological and local streamline based features are good at detecting and describing features of the resulting flow, but not for features hidden or moved by superposition. The resulting errors in the analysis of the feature parameters are high in the case of superposition, even resulting in not detecting a vortex at all. Vorticity and template matching clearly depict the vortices in the resulting scalar fields, but even these results can be influenced by superposition. Assimilation effects, like two vortices with same parameters but different rotation direction which cancel each other out completely, effect the results of feature detection using any feature definition. Comparing the definitions for their robustness to these annihilation effects, template matching using vector valued masks gives the best results as it is tuned to the projection of the features to the abstract model as described in Section 3.1.

These effects should be considered when using topology or other methods for the visualization for vector fields. Again, though the superposition perception of vector fields can be appropriate for some applications, for others the actual streamlines are of higher importance. There, streamline based approaches for visualizing the flow, like topological methods, are better at

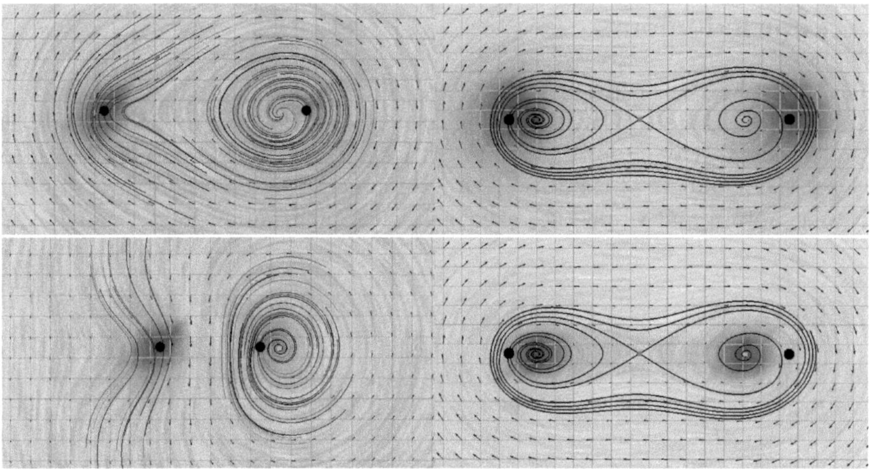

Fig. 5. Superposition and interaction of two Vatistas vortices [17]. All images: The original vortex centers are displayed as black dots. Grid (green), hedgehogs (black arrows), color coding of similarity to a 3x3 rotational mask from high negative values (blue) to high positive values (red). Left: The stronger vortex hides the weaker vortex in streamline based visualizations. Topology only detects one center each (green dot), therefore some streamlines are added. Top right: Template matching detects the true vortex centers. Bottom right: Setting all velocity magnitudes in the field to one, and matching afterwards, yields results more similar to the topological features (coloplate on p. 211).

describing the flow than velocity and direction based feature models like the Vatistas vortex.

5 Conclusion

In this paper, the authors have investigated the borders of hitherto existing topological methods on flow fields. Some of the aspects discussed here may explain why topology is not used more often by engineers. Engineers often think in terms of velocity, vorticity, and resulting feature models like the Vatistas vortex which are more tuned to a superposition perspective of the flow field. Parameters of the feature models thus can often not be determined or visualized using topological methods. Methods based on other feature definitions like pattern matching approaches have to be used then.

Nevertheless, topology is a useful tool for a first overview of and a detailed look at a data set when streamlines or the actual flow is to be studied. Furthermore, when analyzing or visualizing a data set, usually more than one method is used.

Acknowledgments

We thank Markus Rütten, DLR Göttingen, for the Kármán vortex street. Further thanks go to Dr. Berend van der Wall, DLR Braunschweig, and the whole HART II team for data of the HART II test. Furthermore, we like to thank the members of the FAnToM development team at the Universities of Kaiserslautern and Leipzig for their help with programming.

References

1. C. L. Burley, T. F. Brooks, B. G. van der Wall, H. Richard, M. Raffel, P. Beaumier, Y. Delrieux, J. W. Lim, Y. H. Yu, C. Tung, K. Pengel, and E. Mercker. Rotor wake vortex definition - initial evaluation of results of the HART-II study. In *28th EUROPEAN ROTORCRAFT FORUM*, Bristol, England, 2002.
2. J. Ebling and G. Scheuermann. Clifford convolution and pattern matching on vector fields. In *Proceedings of IEEE Visualization 2003*, pages 193–200, Los Alamitos, CA, 2003. IEEE Computer Society.
3. J. Ebling and G. Scheuermann. Fourier transform on multivector fields. *IEEE Transactions on Visualization and Computer Graphics*, 11(4):469–479, 2005.
4. J. Ebling, G. Scheuermann, and B.G. van der Wall. Analysis and visualization of 3-c piv images from HART II using image processing methods. In *Data Visualization 2005*, pages 161–168, 2005.
5. E. B. Heiberg, T. Ebbers, L. Wigström, and M. Karlsson. Three dimensional flow characterization using vector pattern matching. *IEEE Transactions on Visualization and Computer Graphics*, 9(3):313–319, 2003.
6. J. Jeong and F. Hussain. On the identification of a vortex. *Journal of Fluid Mechanics*, 285:69 – 94, 1995.
7. K. Polthier and E. Preuß. Variational approach to vector field decomposition. In *Data Visualization 2000*, pages 113–134, 2000.
8. K. Polthier and E. Preuß. Identifying vector field singularities using a discrete hodge decomposition. *Visualization and Mathematics III*, pages 775–792, 2003.
9. F. H. Post, B. Vrolijk, H. Hauser, R. S. Laramee, and H. Doleitsch. The state of the art in flow visualization: Feature extraction and tracking. *Computer Graphics Forum*, 22(4):775–792, 2003.
10. W.J.M. Rankine. *Manual of Applied Mechanics.* C. Griffen Co., London, 1858.
11. M. Roth. *Automatic Extraction Of Vortex Core Lines And Other Line Type Features For Scientific Visualization.* PhD thesis, ETH, 2000.
12. M.P. Scully. Computation of helicopter rotor wake geometry and it's influence on rotor harmonic airloads. Technical Report ASRL TR 178-1, Masachussetts Institute of Technology, March 1975.
13. B. Shneiderman. The eyes have it: A task by data type taxonomy for information visualizations. In *In Proceedings of the IEEE Symposium on Visual Languages*, pages 336–343, Washington, 1996. IEEE Computer Society Press.
14. Y. Tong, S. Lombeyda, A. Hirani, and M. Desbrun. Discrete multiscale vector field decomposition. *ACM Transactions on Graphics*, 22(3):445–452, 2003.
15. T. Urness, V. Interante, E. Logmire, I. Marusic, and B. Ganapathisubramani. Interactive poster: Illustrating different convection velocities of turbulent flow.

In *Poster Compendium of IEEE Visualization 2004*, Los Alamitos, CA, 2004. IEEE Computer Society Press.
16. B. G. van der Wall, C. L. Burley, Y. H. Yu, H. Richard, K. Pengel, and P. Beaumier. The HART II test - measurement of helicopter rotor wakes. *Aerospace Science and Technology*, 8(4):273–284, 2004.
17. G.H. Vatistas. New model for intense self-similar vortices. *Experiments in Fluids*, 14(4):462–469, 1998.
18. A. Wiebel, C. Garth, and G. Scheuermann. Localized flow analysis of 2d and 3d vector fields. In *Data Visualization 2005*, pages 143–150. Eurographics Association, 2005.

On the Applicability of Topological Methods for Complex Flow Data

Holger Theisel[1], Tino Weinkauf[2], Hans-Christian Hege[2], and Hans-Peter Seidel[3]

[1] Bielefeld University, Germany
theisel@techfak.uni-bielefeld.de
[2] Zuse Institute Berlin (ZIB), Germany
{weinkauf,hege}@zib.de
[3] MPI Informatik Saarbrücken, Germany
hpseidel@mpi-inf.mpg.de

Summary. In this paper we study the applicability of topological methods for creating expressive, feature revealing visualizations of 3D vector fields. 3D vector fields can become very complex by having a high number of critical points and separatrices. Moreover, they may have a very sparse topology due to a small number of critical points or their total absence. We show that classical topological methods based on the extraction of separation surfaces are poorly suited for creating expressive visualizations of topologically complex fields. We show this fact by pointing out that the number of sectors of different flow behavior grows quadratically with the number of critical points – contrary to 2D vector fields. Although this limits the applicability of topological methods to a certain degree, we demonstrate the extensibility of this limit by using further simplifying methods like saddle connectors. For 3D vector fields with a very sparse topology, topological visualizations may fail to reveal the features inherent to the field. We show how to overcome this problem for a certain class of flow fields by removing the ambient part of the flow.

1 Introduction

Topological methods are standard tools to visualizing 2D vector fields. They gained a rather high popularity because they offer to express even a complex flow behavior by only a limited number of graphical primitives. The main idea behind them is to segment the vector field into areas of different flow behavior. To do so, so-called separatrices, mainly starting from critical points, are extracted and visualized.

Topological methods for 2D vector fields have been introduced to the visualization community in [11]. Later they were extended to higher order critical points [23], boundary switch points [6], and closed separatrices [34]. In addition, topological methods have been applied to simplify [6, 7, 28, 29], smooth

[33], compress [17, 25, 16] and construct [24, 32] vector fields as well as to compute distance functions of them [15, 2, 26].

For 3D vector fields, there is a remarkable gap between the need for simplified visual representations of vector fields, the knowledge of topological concepts, and their application as a visual analysis tool. The need for simplified visual representations of 3D vector fields is even higher than for 2D vector fields, since 3D vector fields tend to have a significantly higher amount of information to be visualized. Also, topological concepts for 3D vector fields are well-understood. Nevertheless, there are only a few applications of topological methods for 3D vector fields. Similar to 2D vector fields, [12] proposed methods for detecting and classifying first order critical points by an eigenvalue/eigenvector analysis of the Jacobian matrix. A system for visualizing the topological skeleton of 3D vector fields has been presented in [9]. Topological skeletons of particular analytic 3D vector fields are extracted in [18, 10]. Mahrous et. al [20, 19] obtain a topological segmentation of a vector field by densely sampling stream lines over the field and clustering areas where a similar inflow/outflow behavior of the stream lines is observed. [27] and [31] extract and visualize the intersection curves of the separation surfaces to obtain less cluttered and more expressive visualizations.

Like every visualization technique, topological methods do not give expressive visualizations for all kinds of 3D vector data. In fact, topological methods are limited to rather moderate topological complexity which becomes manifest in the number of present topological features: if only very few features are present (or no features at all), topological methods fail. On the other hand, if there are too many topological features, topological methods fail as well because they produce cluttered visualizations which are hard (or even impossible) to interpret.

It is the purpose of this paper to study where the limits for applying topological methods are, and to present solutions to extend these limits in both directions. For the upper limit (i.e. the fact that topological methods fail if the data is too complex), a number of technical and perceptional reasons are known. Here we show that there is an additional theoretical reason which strongly limits 3D topology to rather simple data sets. This reason lies in the fast growing number of sectors of different flow behavior. We show that – contrary to 2D vector fields – the number of sectors of different flow behavior grows in the worst case quadratically with the number of present topological features (i.e. critical points). As a consequence of this, classical topological methods (focusing on extracting critical points and separation surfaces) are not relevant for topologically complex vector fields. Nevertheless, we show that for simplifying methods like saddle connectors [27], the upper limit is above the currently considered topological complexity. In fact, we apply topological methods to topologically far more complex vector fields than previously considered in the visualization community.

For the lower limit of topological methods (i.e. the fact that topological methods fail for a very poor topology), a simple and well-known solution to move the topological complexity up to a range where expressive visualizations are possible is to remove the ambient part of the flow. We show that this approach can reveal important structures of certain types of flow.

The rest of the paper is organized as follows: section 2 recollects topological concepts for 3D vector fields and their visualization. Section 3 studies the upper limit of topological methods by counting the number of sectors of different flow behavior. Section 4 demonstrates at an example how topological methods can be applied to topologically more complex data sets. Section 5 presents and discusses a solution for dealing with data sets of a very low topological complexity. Section 6 draws conclusions.

2 3D Vector Field Topology and its Visualization

Topological structures of 3D vector fields are well-understood in the visualization community for many years [12, 1, 3, 21]. In this section, we collect the most important concepts and properties, and we review approaches to visualizing them.

2.1 Critical Points

Consider a 3D vector field

$$\mathbf{v}(x,y,z) = \begin{pmatrix} u(x,y,z) \\ v(x,y,z) \\ w(x,y,z) \end{pmatrix}. \quad (1)$$

A first order critical point $\mathbf{x}_0$ (i.e., $\mathbf{v}(\mathbf{x}_0) = \mathbf{0}$) can be classified by an eigenvalue/eigenvector analysis of the Jacobian matrix $\mathbf{J}_\mathbf{v}(\mathbf{x}) = \nabla\mathbf{v}(\mathbf{x})$, iff $\det(\mathbf{J}_\mathbf{v}(\mathbf{x}_0)) \neq 0$. Let $\lambda_1, \lambda_2, \lambda_3$ be the eigenvalues of $\mathbf{J}_\mathbf{v}(\mathbf{x}_0)$ ordered according to their real parts, i.e. $Re(\lambda_1) \leq Re(\lambda_2) \leq Re(\lambda_3)$. Furthermore, let $\mathbf{e}_1, \mathbf{e}_2, \mathbf{e}_3$ be the corresponding eigenvectors. The sign of the real part of an eigenvalue λ_i denotes – together with the corresponding eigenvector $\mathbf{e}_i$ – the flow direction: Positive values represent an *outflow* and negative values an *inflow* behavior. This leads to the following classification of first order critical points:

Sources:	$0 < Re(\lambda_1) \leq Re(\lambda_2) \leq Re(\lambda_3)$
Repelling saddles:	$Re(\lambda_1) < 0 < Re(\lambda_2) \leq Re(\lambda_3)$
Attracting saddles:	$Re(\lambda_1) \leq Re(\lambda_2) < 0 < Re(\lambda_3)$
Sinks:	$Re(\lambda_1) \leq Re(\lambda_2) \leq Re(\lambda_3) < 0$

Thus, sources and sinks consist of a complete outflow/inflow, while saddles have a mixture of both. A repelling saddle has one direction of inflow behavior (called *inflow direction*) and a plane in which a 2D outflow behavior occurs (called *outflow plane*). Similar to this, an attracting saddle consists of an *outflow direction* and an *inflow plane.*

Each of the 4 classes above can be further divided into two stable subclasses by deciding whether or not imaginary parts in two of the eigenvalues are present ($\lambda_1, \lambda_2, \lambda_3$ are not ordered):

Foci:	$Im(\lambda_1) = 0$ and $Im(\lambda_2) = -Im(\lambda_3) \neq 0$
Nodes:	$Im(\lambda_1) = Im(\lambda_2) = Im(\lambda_3) = 0$

An iconic representation is an appropriate visualization for critical points, since vector fields usually contain a finite number of them. Several icons have been proposed in the literature, see [12, 9, 18, 10, 27]. Here we follow the approach of [31] and color the icons depending on the flow behavior: Attracting parts (inflow) are colored blue, while repelling parts (outflow) are colored red. Higher order critical points are not considered in this paper.

2.2 Separatrices

Separatrices are stream lines or stream surfaces which separate regions of different flow behavior. Different kinds of separatrices are possible: They can emanate from critical points or boundary switch curves, or they are closed separatrices without a specific emanating structure. However, in this paper we consider separatrices starting from critical points only.

Due to the homogeneous flow behavior around sources and sinks (either a complete outflow or inflow), they do not contribute to separatrices. Each saddle point creates two separatrices: Considering a repelling saddle $\mathbf{x}_R$, it creates one separation curve (which is a stream line starting in $\mathbf{x}_R$ in the inflow direction by backward integration) and a separation surface (which is a stream surface starting in the outflow plane by forward integration). A similar statement holds for attracting saddles. Since for the segmentation of a vector field into sectors of different flow behavior only the separation surfaces (and not the separation lines) contribute, we only consider separation surfaces in the following.

Contrary to the 2D case, separatrices of 3D vector fields can intersect in a number of stream lines called saddle connectors [27]. Saddle connectors start in the repelling plane of a repelling saddle and end in the attracting plane of an attracting saddle. [31] extends the concept of saddle connectors to boundary switch connectors which are the intersections of separatrices emanating from boundary switch curves.

3 Counting the Number of Sectors

We start with an analysis of existing 3D topological visualization approaches and consider the topological complexity of the treated data sets. Table 1 gives a collection of these techniques, including the topological complexity of the treated examples. We express this complexity by counting the number of critical points, boundary switch curves and separatrices which are present in the application. Table 1 does not intend to give an evaluation of the considered techniques because they focus on different data sets or incorporate other visualization techniques as well. However, table 1 reveals that most of the applications deal only with vector fields of a very low topological complexity. Only the recent papers [27] and [31] consider fairly complex data sets but conclude that classical topological methods are not appropriate there.

reference	#cp	#scp	#bsc	#sbsc
[12]	$\approx$20	$\approx$5	0	0
[9]	$\approx$2	0	0	0
[18]	3	2	0	0
[20]	0	0	$\approx$10	$\approx$10
[19]	1	1	$\approx$10	$\approx$10
[27]	184	121	0	0
[31]	184	121	13	22
[this paper]	452	452	0	0

Table 1. 3D topological visualization approaches and their number of treated topological features; #cp: number of critical points; #scp: number of separatrices starting from critical points; #bsc: number of boundary switch curves; #sbsc: number of separatrices starting from boundary switch curves.

We search for reasons why up to now topological methods have been applied only to rather simple data sets. Two classes of reasons are already known [27]:

1. *Technical reason:* 3D topological methods involve the integration of stream surfaces which is computationally more involved, less stable, and less accurate than the integration of stream lines in 2D.
2. *Perceptional reason:* The sectors of different flow behavior may have a complicated shape and hide each other, making a visual analysis of them a cumbersome task. Figure 6 shows an example of a vector field consisting of 4 saddles which create 6 sectors of different flow behavior. Even for this rather low number of sectors we observe the hiding effect making it hard to distinguish the different sectors.

In recent years the first problem became more and more unimportant due to the dramatic increase of computing capacities and a number of new algorithmic solutions ([13], [8], [22], [30]). One solution for the second problem is the saddle connector approach [27] [31].

Now we show that there is a third reason that topological methods are limited to low-complexity vector fields. We show that – simply spoken – the number of sectors of different flow behavior grows fast when the topological complexity of the vector field increases. As a measure of topological complexity, we take the number of present saddle points. Since most considered data sets have a global index around zero, the number of saddles is approximately half the number of critical points. Then we get a

3. *Theoretical reason:* The number of sectors of different flow behavior grows in the worst case quadratically with the number of saddle points in a 3D vector field.

We show that this reason is a serious limitation of applying topological methods to 3D vector fields. To prove this reason, we present formulas to compute the number of sectors of different flow behavior. To see the differences, we do so both for 2D and 3D vector fields.

3.1 Sector counting for 2D vector fields

2D vector fields generally consist of sources, sinks and saddles where a saddle creates 4 separation curves [11]. We get

Property 1. Given a 2D vector field $\mathbf{v}_{n_S}$ consisting of n_S saddle points, the number $sec(\mathbf{v}_{n_S})$ of sectors of different flow behavior fulfills

$$sec(\mathbf{v}_{n_S}) \leq 3\,n_s + 1 \tag{2}$$

where the equality in (2) can be reached.

Property 1 essentially says that the number of sectors grows linearly with the number of saddle points. To show it, we start with a vector field consisting of only one saddle, as shown in figure 1(a). This saddle divides the domain into four sectors. Now we insert a new saddle as shown in figure 1(b). Since the different sectors are separated by stream lines which must not intersect each other, a new saddle replaces one of the old sectors by 4 new sectors, thus increasing the total number of sectors by 3. This gives

$$sec(\mathbf{v}_{n_S+1}) \leq sec(\mathbf{v}_{n_S}) + 3, \tag{3}$$

which is an inequality since separatrices may end in the same source/sink which reduces the number of sectors. Figure 1(c) illustrates this. (3) and $sec(\mathbf{v}_0) = 1$ gives (2). To complete the proof, we only have to show that the equality in (2) can be reached. To do so, we construct a vector field $\mathbf{v}_{n_S}$ with $sec(\mathbf{v}_{n_S}) = 3\,n_s + 1$. Figure 2 illustrates the construction of such a simple vector field.

Property 1 gives a reason why 2D topological methods are rather popular even for fairly complex vector fields: the number of sectors to be distinguished grows only slowly (in fact linearly) with the increasing of the topological complexity (i.e. the number of saddle points). As we will show now, this does not hold for 3D vector fields.

3.2 Sector counting for 3D vector fields

For 3D vector fields, the number of sectors of different flow behavior depends in the worst case quadratically on the number of saddle points. We formulate

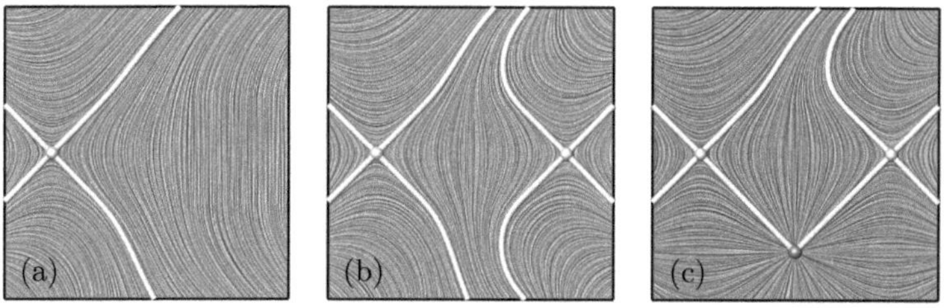

Fig. 1. (a) a single saddle point segments the domain into 4 sectors; (b) an additionally included saddle increases the total number of sectors by 3; (c) if separatrices end in the same source/sink, two sectors are merged.

Fig. 2. Vector field $\mathbf{v}_{n_S}$ with $sec(\mathbf{v}_{n_S}) = 3\,n_s + 1$

Property 2. Given a 3D vector field $\mathbf{v}_{n_R,\,n_A}$ consisting of n_R repelling saddles and n_A attracting saddles, for the number $sec(\mathbf{v}_{n_R,\,n_A})$ of sectors of different flow behavior the inequality holds

$$sec(\mathbf{v}_{n_R,\,n_A}) \;\leq\; (n_R + 1)(n_A + 1) \tag{4}$$

where the equality in (4) can be reached.

To show property 2, we start with a simple vector field $\mathbf{v}_{1,0}$ consisting only of one repelling saddle $\mathbf{x}_R$, as shown in figure 3(a). The separation surface created by $\mathbf{x}_R$ divides $\mathbf{v}_{1,0}$ into two sectors. If we insert a new saddle, this can be either an attracting saddle $\mathbf{y}_A$ or a repelling saddle $\mathbf{y}_R$ as well. In the last case, the separation surfaces of $\mathbf{x}_R$ and $\mathbf{y}_R$ create three sectors of different flow behavior since they must not intersect. In case of a newly inserted attracting saddle $\mathbf{y}_A$, two cases are possible:

- The separation surface of $\mathbf{y}_A$ does not intersect the separation surface of $\mathbf{x}_R$. In this case, one of the old sectors is divided into two new sectors, and the total number of sectors is increased by 1. Figure 3(b) gives an illustration.
- The separation surface of $\mathbf{y}_A$ intersects the separation surface of $\mathbf{x}_R$. In this case, each of the two old sectors is divided into two new sectors. Thus, the total number of sectors is increased by 2. Figure 3(c) illustrates this.

As we can see from the simple example above, the total number of sectors does not only depend on the number of saddles but also on the number of saddle connectors.

For now we assume that every repelling saddle has a connector to every attracting saddle. Given a vector field $\mathbf{v}_{n_R,\,n_A}$, we consider the insertion of

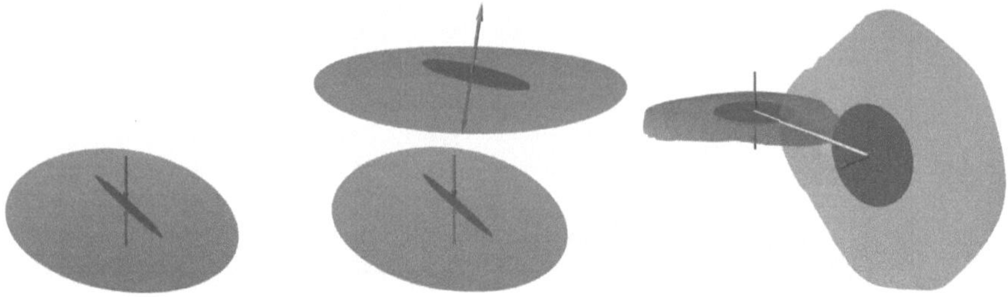

(a) Simple vector field with one $\mathbf{x}_R$: two sectors are present.

(b) Inserting $\mathbf{y}_A$ *without* connector gives 3 sectors.

(c) Inserting $\mathbf{y}_A$ *with* connector gives 4 sectors.

Fig. 3. Correlation between number of sectors, saddles and connectors.

a new attracting saddle $\mathbf{x}_A$. Assuming that $\mathbf{x}_A$ has a connector to all n_R repelling saddles of $\mathbf{v}_{n_R,\,n_A}$, $\mathbf{x}_A$ divides n_R+1 of the old sectors into two new sectors each. We get

$$sec(\mathbf{v}_{n_R,\,n_A+1}) \leq sec(\mathbf{v}_{n_R,\,n_A}) + n_R + 1 \tag{5}$$

and in a similar way

$$sec(\mathbf{v}_{n_R+1,\,n_A}) \leq sec(\mathbf{v}_{n_R,\,n_A}) + n_A + 1. \tag{6}$$

(5), (6) and $sec(\mathbf{v}_{0,0}) = 1$ give (4).

To complete the proof of property 2, we construct an example vector field $\mathbf{v}_{n_R,\,n_A+1}$ with $sec(\mathbf{v}_{n_R,\,n_A}) = (n_R+1)(n_A+1)$. To do so, we use the topological vector field construction approach described in [32]. We place the n_R repelling saddles to the locations $(1, \frac{n_A}{2}, -d)$, $(2, \frac{n_A}{2}, -d)$, ..., $(n_R, \frac{n_A}{2}, -d)$ in such a way that the inflow plane of each saddle is parallel to the $y-z$ plane of the underlying Euclidian coordinate system. Furthermore, we place n_R+1 sources at the locations $(0.5, \frac{n_A}{2}, -d)$, $(1.5, \frac{n_A}{2}, -d)$, ..., $(n_R+0.5, \frac{n_A}{2}, -d)$. To place the n_A attracting saddles, we choose the locations $(\frac{n_R}{2}, 1, d)$, $(\frac{n_R}{2}, 2, d)$, ..., $(\frac{n_R}{2}, n_A, d)$. In addition we place n_A+1 sinks at the locations $(\frac{n_R}{2}, 0.5, d)$, $(\frac{n_R}{2}, 1.5, d)$, ..., $(\frac{n_R}{2}, n_A+0.5, d)$. The positive number d describes the distance of the two rows of saddles. Figure 4a illustrates the location of the critical points for the example $\mathbf{v}_{4,4}$.

In the next step we have to construct a system of connectors such that each repelling saddle is connected to each attracting saddle. This is a set of $n_R \cdot n_A$ curves which must not intersect each other (except in the saddles themselves). Given the arrangement of critical points described above, this can easily be done as illustrated in figure 4(a) for $\mathbf{v}_{4,4}$. The complete constructed vector field ensures that for any source $\mathbf{x}_{So}$ and for any sink $\mathbf{x}_{Si}$ there are stream lines starting in $\mathbf{x}_{So}$ and ending in $\mathbf{x}_{Si}$. Figure 4(b) shows the complete topological skeleton of $\mathbf{v}_{4,4}$. Figure 4(c) shows an example of the constructed vector field $\mathbf{v}_{20,20}$ consisting of 441 sectors of different flow behavior.

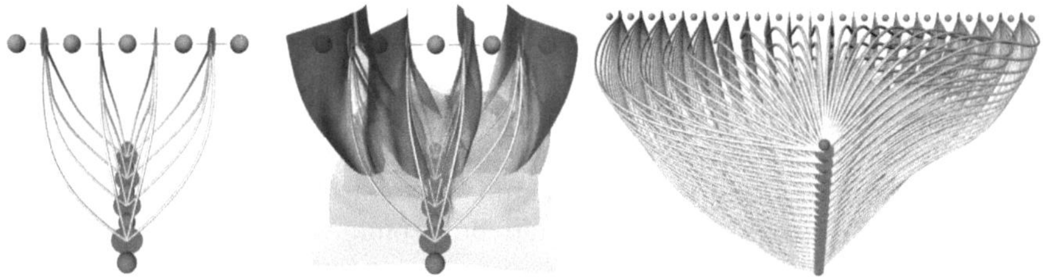

(a) $\mathbf{v}_{4,4}$: critical points and 16 saddle connectors. (b) $\mathbf{v}_{4,4}$: separation surfaces. (c) $\mathbf{v}_{20,20}$: critical points and 400 saddle connectors.

Fig. 4. Constructed vector fields $\mathbf{v}_{4,4}$ and $\mathbf{v}_{20,20}$.

Property 2 can be concretized by incorporating not only the number of critical points but also the number of connectors:

Property 3. Given a 3D vector field $\mathbf{v}_{n_R,\, n_A,\, n_{Co}}$ consisting of n_R repelling saddles, n_A attracting saddles and n_{Co} saddle connectors, for the number $sec(\mathbf{v}_{n_R,\, n_A,\, n_{Co}})$ of sectors of different flow behavior the inequality holds

$$sec(\mathbf{v}_{n_R,\, n_A,\, n_{Co}}) \leq n_R + n_A + n_{Co} + 1. \tag{7}$$

To show property 3, we assume a vector field $\mathbf{v}_{n_R,\, n_A,\, n_{Co}}$ in which we insert a new attracting saddle $\mathbf{x}_A$. Further we assume that $\mathbf{x}_A$ creates m new saddle connectors, i.e. $\mathbf{x}_A$ is connected with m repelling saddles of $\mathbf{v}_{n_R,\, n_A,\, n_{Co}}$. In this case, $m+1$ sectors of the old vector field are divided into two new sectors each. We obtain

$$sec(\mathbf{v}_{n_R,\, n_A+1,\, n_{Co}+m}) \leq sec(\mathbf{v}_{n_R,\, n_A,\, n_{Co}}) + m + 1 \tag{8}$$

and in a similar way

$$sec(\mathbf{v}_{n_R+1,\, n_A,\, n_{Co}+m}) \leq sec(\mathbf{v}_{n_R,\, n_A,\, n_{Co}}) + m + 1. \tag{9}$$

This and $sec(\mathbf{v}_{0,0,0}) = 1$ gives (7).

Remarks:

1. The conditions in properties 2 and 3 are formulated as inequality because - similar to the 2D case - separatrices might end in the same critical points which leads to a reduction of the total number of sectors.
2. Properties 2 and 3 did not consider separatrices emanating from boundary switch curves. However, their quantitative behavior is similar to the separatrices from saddle connectors: the number of sectors grows in the worst case quadratically to the number of boundary switch curves.
3. Property 3 shows that in the best case the number of sectors grows linearly with the number of saddles. This happens if no saddle connectors exist at all. However, the example of the data sets in section 4 show that a higher number of saddle connectors usually exists.

4. Properties 2 and 3 considered at most one connector between a repelling saddle $\mathbf{x}_R$ and an attracting saddle $\mathbf{x}_A$. If multiple connectors are present, a sector (describing the flow from one particular source to one particular sink) may consist of different unconnected parts.
5. The sector counting presented here is a worst case estimation. Although an average case estimation would be useful, we are not aware of any approaches for this.

3.3 Interpretation of sector counting

From the sector counting approach in section 3 we draw the conclusion that classical 3D topological methods are limited to topologically rather simple vector fields. If the topological complexity (i.e. the number of saddles) grows, the number of sectors of different flow behavior very soon exceeds the limit of what can be distinguished in visualization. The only solution for this is to apply simplifying topological methods. The topological skeleton may be simplified by removing unimportant critical points or collapse clusters of critical points to a higher order one. While these methods are well-established for 2D vector fields ([6] [29]), we are not aware of any 3D extensions. The simplifying method we consider here are saddle connectors which we apply to more complex data sets in the next section.

4 Topologically Rich Vector Fields

From examples in [27] and [31] and from section 3 it is known that topological methods are hardly applicable for topologically complex vector fields. However, in this section we investigate topological methods for a 3D vector field of such a high topological complexity as it has – to the best of our knowledge – not been treated in the literature yet.

Figure 5 shows the transitional flow around a backward-facing step. The flow field is obtained from a numerical simulation of Kaltenbach and Janke at a Reynolds number of Re_H=3000 based on oncoming velocity and on step height. The corresponding boundary conditions are described in [14]. The data set contains 452 critical points which are visualized in figure 5(a). Since the vector field is divergence-free, all of them are saddles.

A complete extraction and visualization of the separation surfaces was possible only at a very coarse resolution (figure 5(b)) since for higher resolutions the number of produced triangles very soon exceeded the limits of our available hardware. But even if we were able to process the complete skeleton at a high resolution, already figure 5(a) shows that an expressive topological visualization can not be achieved due to the sheer number of surfaces. Nevertheless, we were able to extract the intersection curves of these surfaces, namely the saddle connectors. As known from [27], this algorithm exhibits far less memory consumption for both extraction and display than the treatment of separation surfaces themselves.

Figure 5(d) shows the visualization of the 1023 extracted saddle connectors. As we can see there, certain flow structures become visible: there are 3 rather independent clusters of turbulent flow behavior. However, it also shows that the segmentation property gets lost for vector fields of this complexity: no regions of different flow behavior can be distinguished any more from such a large number of saddle connectors. In fact, for complex data sets the approach of saddle connectors shifts from a topological segmentation techniques to a stream line selection technique.

Even if the complexity is too rich for a direct topological visualization, topological information can be used to parameterize other visualization techniques. We demonstrate this in figure 5(e). Here, we made use of the extracted critical points and seeded stream lines close to them. This gives a far more expressive visualization than figure 5(c), where stream lines have been seeded homogeneously over the whole domain. Both figures show the same number of stream lines.

Figure 5(e) illuminates the coherent structures of this type of flow: The flow separates at the corner of the step. The resulting shear layer rolls up in two Kelvin-Helmholtz vortices. In the downstream direction, the streamlines form bundles due to secondary streamwise vorticity. The fluid experiences a small backward flow in the upstream region below the shear layer.

5 Topologically Sparse Vector Fields

While dealing with a variety of data sets, we encountered vector fields where topological methods totally fail due to the absence of critical points and boundary switch curves. While it might not be possible to overcome this problem for all kinds of data, there is a solution for an important class of flow fields that exhibit a constant ambient flow part: all convections, i.e. coherent structures, move with nearly the same velocity and direction inside the flow. Their corresponding topological structures cannot be extracted since the ambient flow part cancels out the critical points. This clearly shows the Galilean-variance of topological examinations. By subtracting the ambient flow part, i.e. choosing a certain frame of reference, the coherent structures become visible using topological methods. To ensure meaningful results, this manipulation must be motivated by the physical interpretation of the data.

Consider the mixing layer visualized in figure 7(a), where the flow moves downstream in both layers and the magnitude of the upper layer is three times larger than in the lower layer. The data set has been computed with a pseudo-spectral direct numerical simulation employing the computational domain and boundary conditions of Comte, Silvestrini & Bégou [4]. The Reynolds number is 100 based on the initial shear-layer thickness and convection velocity.

No critical points are present in this original frame of reference. In figure 7(b) we have chosen to subtract the constant vector field $(1,0,0)^T$. This yields the frame of reference where the flow in both layers has the same magnitude, but a different direction. The physical interpretation behind this manipulation is that we move as a observer with the same velocity and direction as the convections. The topology of this frame clearly shows formations of focus saddles indicating Kelvin-Helmholtz vortices, which alternate with formations of node saddles.

This example shows that topological methods yield expressive visualizations even for initially topologically sparse vector fields if a frame of reference can be chosen with regards to physical interpretation. Nevertheless, Galilean-invariant methods, like e.g. visualization of vortex regions, overcome the problem of finding the "right" frame of reference.

6 Conclusions

In this paper we made the following contributions:

- We have shown that – contrary to the 2D case – for 3D vector fields the number of sectors of different flow behavior grows in the worst case quadratically to the number of saddle points.
- We applied topological methods to more complex 3D vector fields than previously done in the literature.
- We have discussed that for some flow data of poor topological complexity, a removal of the ambient flow makes topological methods applicable.

We conclude that classical topological methods for 3D vector fields including critical points and separation surfaces are only of rather low relevance for most practical data sets. In fact, without simplifying methods like critical point clustering, critical point removing or saddle connectors, 3D topological methods won't get such a popularity as for 2D vector fields.

For saddle connectors we have shown that they are applicable to rather complex data, but from a certain complexity on the segmentation property gets lost, and saddle connectors are mainly perceived as stream lines.

For the future we expect an ongoing research on topology simplifying techniques because they seem the only promising way to make topological methods applicable to very complex 3D data sets.

Acknowledgements

We thank Bernd R. Noack for the fruitful discussions and supply of simulation data, which was kindly provided by Kaltenbach and Janke (step) as well as Pierre Comte (mixing layer).

All visualizations in this paper have been created using AMIRA – a system for advanced visual data analysis [5] (see `http://amira.zib.de/`).

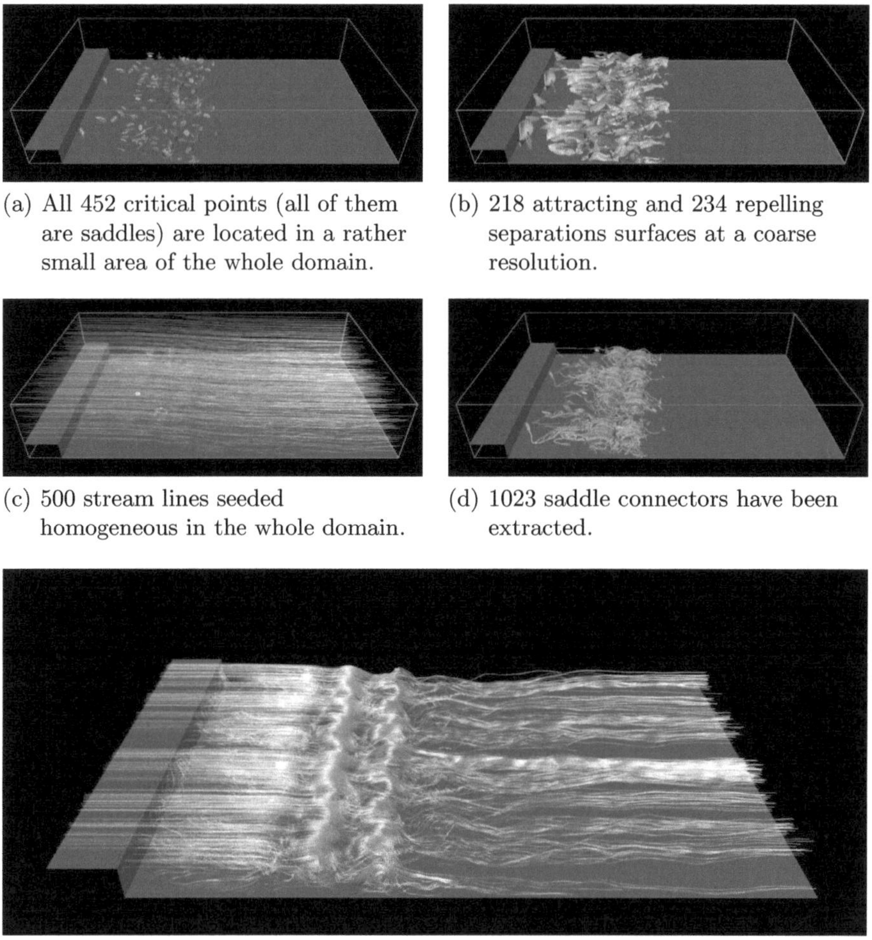

(a) All 452 critical points (all of them are saddles) are located in a rather small area of the whole domain.

(b) 218 attracting and 234 repelling separations surfaces at a coarse resolution.

(c) 500 stream lines seeded homogeneous in the whole domain.

(d) 1023 saddle connectors have been extracted.

(e) 500 stream lines seeded near critical points.

Fig. 5. Flow around a backward-facing step (colorplate on p. 212).

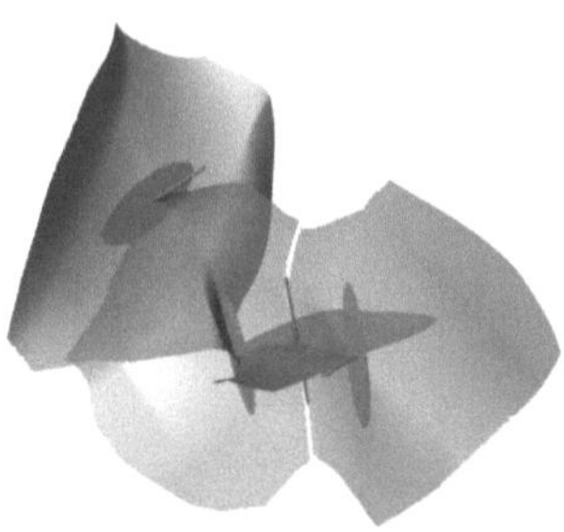

Fig. 6. Simple topological skeleton consisting of 4 saddles; the 6 resulting sectors of different flow behavior can hardly be distinguished (colorplate on p. 212).

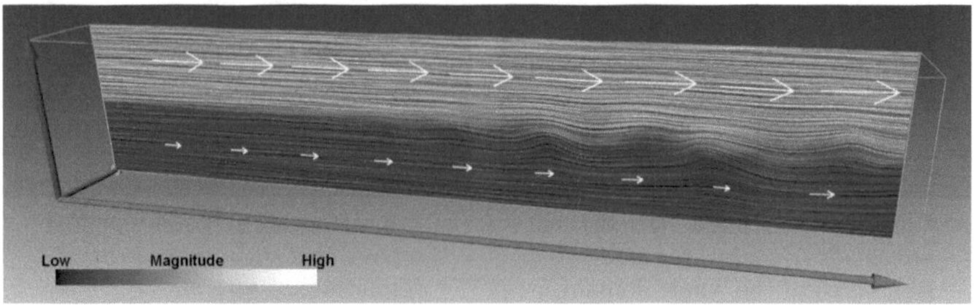

(a) Original frame of reference. No critical points are present.

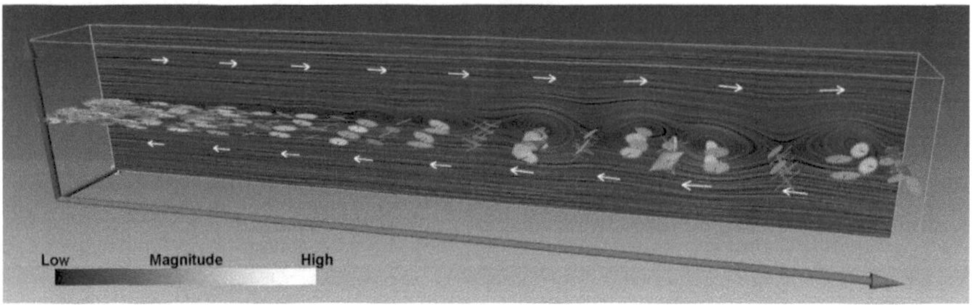

(b) Frame of reference chosen such that both layers have the same magnitude. 348 saddle points have been detected.

Fig. 7. Mixing layer (colorplate on p. 213).

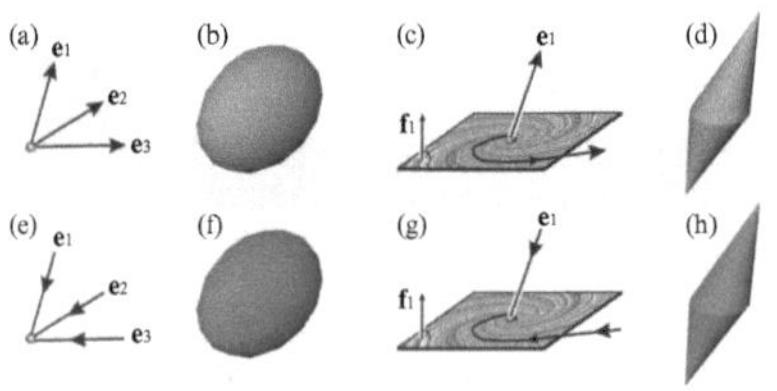

(a) Sources & sinks and their icons.

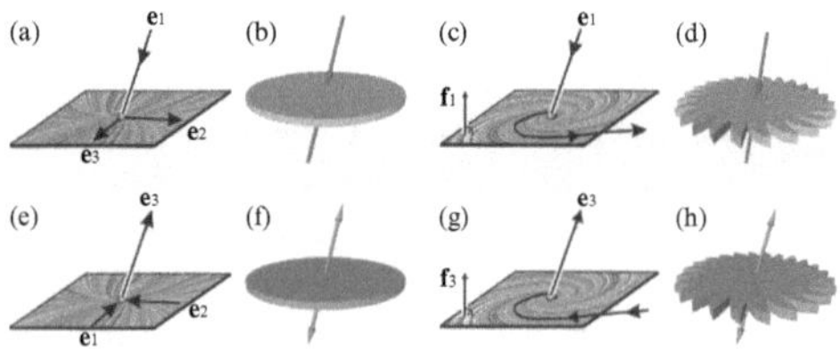

(b) Saddles and their icons.

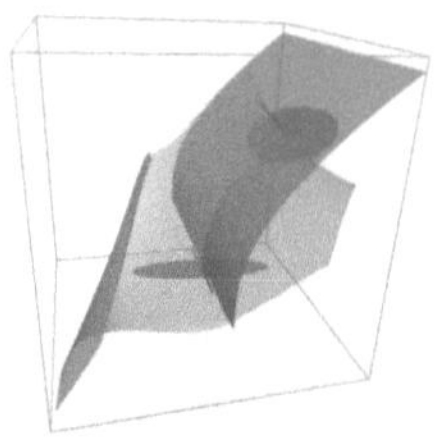

(c) Separation surfaces of the saddles.

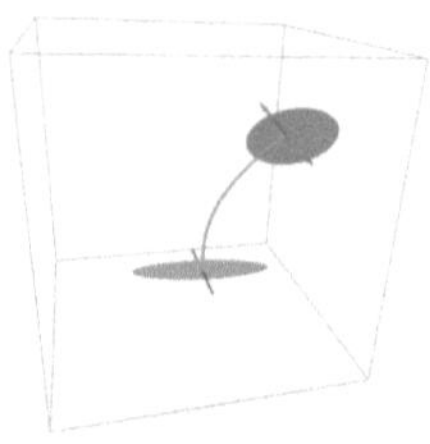

(d) Saddle connector.

Fig. 8. Critical points and definition of saddle connectors (colorplate on p. 213).

References

1. D. Asimov. Notes on the topology of vector fields and flows. Technical report, NASA Ames Research Center, 1993. RNR-93-003.
2. R. Batra, K. Kling, and L. Hesselink. Topology based vector field comparison using graph methods. In *Proc. IEEE Visualization '99, Late Breaking Hot Topics*, pages 25–28, 1999.
3. M. S. Chong, A. E. Perry, and B. J. Cantwell. A general classification of three-dimensional flow fields. *Physics of Fluids A*, 2(5):765–777, 1990.
4. P. Comte, J.H. Silvestrini, and P. Bégou. Streamwise vortices in Large-Eddy Simulations of mixing layer. *Eur. J. Mech. B*, 17:615–637, 1998.
5. Stalling D, M. Westerhoff, and H.C. Hege. Amira: A highly interactive system for visual data analysis. *The Visualization Handbook*, pages 749–767, 2005.
6. W. de Leeuw and R. van Liere. Collapsing flow topology using area metrics. In *Proc. IEEE Visualization '99*, pages 149–354, 1999.
7. W. de Leeuw and R. van Liere. Visualization of global flow structures using multiple levels of topology. In *Data Visualization 1999. Proc. VisSym 99*, pages 45–52, 1999.
8. A. Van Gelder. Stream surface generation for fluid flow solutions on curvilinear grids. In *Data Visualization 2001. Proc. VisSym 01*, 2001.
9. A. Globus, C. Levit, and T. Lasinski. A tool for visualizing the topology of three-dimensional vector fields. In *Proc. IEEE Visualization '91*, pages 33–40, 1991.
10. H. Hauser and E. Gröller. Thorough insights by enhanced visualization of flow topology. In *9th international symposium on flow visualization*, 2000.
11. J. Helman and L. Hesselink. Representation and display of vector field topology in fluid flow data sets. *IEEE Computer*, 22(8):27–36, August 1989.
12. J. Helman and L. Hesselink. Visualizing vector field topology in fluid flows. *IEEE Computer Graphics and Applications*, 11:36–46, May 1991.
13. J. Hultquist. Constructing stream surfaces in steady 3D vector fields. In *Proc. IEEE Visualization '92*, pages 171–177, 1992.
14. H.-J. Kaltenbach and G. Janke. Direct numerical simulation of flow separation behind a swept, rearward-facing step at re_H=3000. *Physics of Fluids*, 12:2320–2337, 2000.
15. Y. Lavin, R.K. Batra, and L. Hesselink. Feature comparisons of vector fields using earth mover's distance. In *Proc. IEEE Visualization '98*, pages 103–109, 1998.
16. S. Lodha, N. Faaland, and J. Renteria. Topology preserving top-down compression of 2d vector fields using bintree and triangular quadtrees. *IEEE Transactions on Visualization and Computer Graphics*, 9(4):433–442, 2003.
17. S.K. Lodha, J.C. Renteria, and K.M. Roskin. Topology preserving compression of 2D vector fields. In *Proc. IEEE Visualization 2000*, pages 343–350, 2000.
18. H. Löffelmann, H. Doleisch, and E. Gröller. Visualizing dynamical systems near critical points. In *Spring Conference on Computer Graphics and its Applications*, pages 175–184, Budmerice, Slovakia, 1998.
19. K. Mahrous, J. Bennett, B. Hamann, and K. Joy. Improving topological segmentation of three-dimensional vector fields. In *Data Visualization 2003. Proc. VisSym 03*, pages 203–212, 2003.

20. K. Mahrous, J. Bennett, G. Scheuermann, B. Hamann, and K. Joy. Topological segmentation in three-dimensional vector fields. *IEEE Transactions on Visualization and Computer Graphics*, 10(2):198–205, 2004.
21. P. A. Philippou and R. N. Strickland. Vector field analysis and synthesis using three dimensional phase portraits. *Graphical Models and Image Processing*, 59:446–462, November 1997.
22. G. Scheuermann, T. Bobach, H. Hagen K. Mahrous, B. Hamann, K. Joy, and W. Kollmann. A tetrahedra-based stream surface algorithm. In *Proc. Visualization 01*, pages 151 – 158, 2001.
23. G. Scheuermann, H. Krüger, M. Menzel, and A. Rockwood. Visualizing nonlinear vector field topology. *IEEE Transactions on Visualization and Computer Graphics*, 4(2):109–116, 1998.
24. H. Theisel. Designing 2D vector fields of arbitrary topology. *Computer Graphics Forum (Eurographics 2002)*, 21(3):595–604, 2002.
25. H. Theisel, Ch. Rössl, and H.-P. Seidel. Compression of 2D vector fields under guaranteed topology preservation. *Computer Graphics Forum (Eurographics 2003)*, 22(3):333–342, 2003.
26. H. Theisel and T. Weinkauf. Vector field metrics based on distance measures of first order critical points. In *Journal of WSCG*, volume 10:3, pages 121–128, 2002.
27. H. Theisel, T. Weinkauf, H.-C. Hege, and H.-P. Seidel. Saddle connectors - an approach to visualizing the topological skeleton of complex 3D vector fields. In *Proc. IEEE Visualization 2003*, pages 225–232, 2003.
28. X. Tricoche, G. Scheuermann, and H. Hagen. A topology simplification method for 2D vector fields. In *Proc. IEEE Visualization 2000*, pages 359–366, 2000.
29. X. Tricoche, G. Scheuermann, and H. Hagen. Continuous topology simplification of planar vector fields. In *Proc. Visualization 01*, pages 159 – 166, 2001.
30. J. van Wijk. Implicit stream surfaces. In *Proc. Visualization 93*, pages 245–252, 1993.
31. T. Weinkauf, H. Theisel, H.-C. Hege, and H.-P. Seidel. Boundary switch connectors for topological visualization of complex 3D vector fields. In *Data Visualization 2004. Proc. VisSym 04*, pages 183–192, 2004.
32. T. Weinkauf, H. Theisel, H.-C. Hege, and H.-P. Seidel. Topological construction and visualization of higher order 3D vector fields. *Computer Graphics Forum (Eurographics 2004)*, 23(3):469–478, 2004.
33. R. Westermann, C. Johnson, and T. Ertl. Topology-preserving smoothing of vector fields. *IEEE Transactions on Visualization and Computer Graphics*, 7(3):222–229, 2001.
34. T. Wischgoll and G. Scheuermann. Detection and visualization of closed streamlines in planar flows. *IEEE Transactions on Visualization and Computer Graphics*, 7(2):165–172, 2001.

Extraction and Visualization of Swirl and Tumble Motion from Engine Simulation Data

Christoph Garth[1], Robert S. Laramee[2], Xavier Tricoche[3], Jürgen Schneider[4], and Hans Hagen[1]

[1] Visualization Group, University of Kaiserslautern
{garth,hagen}@informatik.uni-kl.de
[2] Dept. of Computer Science, Univ. of Wales
r.s.laramee@swansea.ac.uk
[3] SCI Institute, University of Utah
tricoche@sci.utah.edu
[4] AVL, Graz, Austria
juergen.schneider@avl.com

Summary. An optimal combustion process within an engine block is central to the performance of many motorized vehicles. Associated with this process are two important patterns of flow: swirl and tumble motion, which optimize the mixing of fluid within each of an engine's cylinders. The simulation data associated with in-cylinder tumble motion within a gas engine, given on an unstructured, time-varying and adaptive resolution CFD grid, demands robust visualization methods that apply to unsteady flow. Good visualizations are necessary to analyze the simulation data of these in-cylinder flows. We present a range of methods including integral, feature-based, and image-based schemes with the goal of extracting and visualizing these two important patterns of motion. We place a strong emphasis on automatic and semi-automatic methods, including topological analysis, that require little or no user input. We make effective use of animation to visualize the time-dependent simulation data. We also describe the challenges and implementation measures necessary in order to apply the presented methods to time-varying, volumetric grids.

1 Introduction

Among the many design goals of combustion engines, the mixing process of fuel and oxygen occupies an important place. If a good mixture can be achieved, the resulting combustion is both clean and efficient, with all the fuel burned and minimal exhaust remaining. In turn, the mixing process strongly depends on the inflow of the fuel and air components into the combustion chamber or cylinder. If the inlet flow generates sufficient kinetic energy during this valve cycle, the resulting turbulence distributes fuel and air optimally in the combustion chamber. For common types of engines, near-optimal flow patterns are actually known and include, among others, so-called swirl and tumble motions. With the general progress of state-of-the-art CFD simulations, the discipline of engine design is made accessible to both numerical simulation and visualization of the resulting datasets, allowing for rapid testing of engine designs.

Laramee et al. [9] took preliminary steps towards the visualization and analysis of in-cylinder flow. Using a combination of texture-based and geometric techniques, they were able to indirectly visualize the key swirl and tumble patterns in two engine simulation datasets. The approaches they used were essentially manual and they did not consider time-dependent flow. It is the aim of this paper to expand on this previous study by applying additional feature-centric visualizations. Here, we

- focus on topological methods and volumetric approaches, namely cutting-plane and boundary topology and direct volume rendering,
- examine hybrid visualizations that combine different techniques,
- emphasize schemes that can extract swirl and tumble characteristics semi-automatically, and
- handle full time-dependent flow on a time-dependent geometry.

With application by engineers in mind, we present a survey of methods that are useful in this context and demonstrate how they can be effectively applied in engine simulation analysis.

Of particular interest are the time-varying nature of the simulation and the interconnection between visualization methods that treat data of different dimensionality (typically boundary vs. volume data). We study to what extent an analysis of the boundary flow permits reliable insight into the volume of the combustion chamber on the presented examples. As an example, we examine the effect that vortices have on the topological structure on the boundary.

We describe the techniques employed and present a critical discussion of the resulting visualizations from an application standpoint. Although the application domain covered in this work is specific, the conclusions reached can be leveraged in many areas of engineering.

The paper is structured as follows. In Section 2, we describe the application that we based our analysis on, namely two important patterns of in-cylinder flow. Section 3 is concerned with the criteria for our choice of methods. We briefly describe the methods and how they contribute to a satisfactory extraction and visualization of swirl and tumble motions. Some of the technical aspects involved in time-varying unstructured grids are detailed as well. Hybrid combinations of methods are examined in Section 5, before we conclude on the presented work in Section 6.

Remark: In our study of the application, we have found that the possibility of interactive and animated viewing of visualization results greatly enhances the comprehension of occurring structures. The reader is therefore referred to the accompanying video [2] in which we demonstrate this.

2 Engine Simulation Data

From a simplified point-of-view, there are two types of ideal flow patterns in an engine cylinder: swirl motion and tumble motion. Both are rotational motions, however, the axis of rotation is different in each case. Depending on the type of engine, one of these patterns is considered optimal because it maximizes mixing of injected fuel and air, resulting in homogeneous combustion.

In this paper, we treat two datasets showing each of these two types of flow patterns (henceforth termed "swirl motion" and "tumble motion"). The basic geometries of the datasets and the respective desired motion patterns are shown in Figure 1. Although they were generated in the same problem context, the simulation datasets differ in a number of ways.

Swirl Motion in a Diesel Engine

This simulation is the result of a the simulation of steady charge flow in a diesel engine, based on a stationary geometry, resulting in a simple and stable flow. The main axis of motion is aligned with the cylinder axis and is constant in time. The spatial resolution of the single timestep is high with a total of 776,000 unstructured cells on an adaptive resolution grid.

Tumble Motion in a Gas Engine

This dataset results from an unsteady simulation of the charge phase of a gas engine (crank angles 380 to 540). As the piston moves down, the cylinder volume increases by an order of magnitude and the fuel-air mixture entering the cylinder is drawn into a gradually developing tumble pattern. The overall

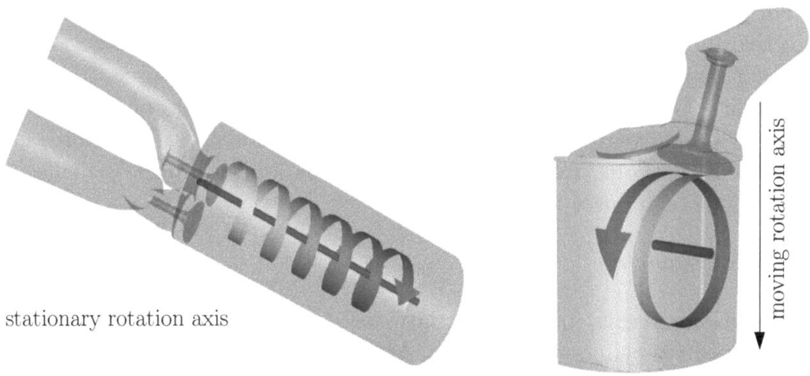

Fig. 1. (Left) Stable, circulating flow pattern in a diesel engine designated as *swirl motion*, with the cylinder axis as the axis of rotation. The flow enters tangentially through the intake ports. (Right) Transient *tumble motion* in a gas engine. The axis of motion moves as the cylinder expands and stays halfway between the top cylinder wall and the piston head at the bottom (not shown).

motion is highly transient and unstable. Both spatial and temporal resolution are relatively low, with the data given on 32 timesteps and the grid consisting of roughly 61,000 unstructured elements at the maximum crank angle. It is interesting to note that the actual mesh topology remains constant throughout all timesteps. This is accomplished by the use of virtual zero-volume cells at the piston head that expand as the piston moves down; only the mesh vertices are changing in time.

Both simulations were computed at the Department of Advanced Simulation Technologies (AST) at AVL (`www.avl.com`) for the design and analysis of specific flow in-cylinder types. The commercially available AVL *Fire* solver was used for the solution of the compressible Navier-Stokes equation with a Finite Volume Method. In addition to the flow vector field, the datasets encompass a number of additional attributes such as temperature and pressure. In this work, we focus on the analysis of the swirl and tumble aspects of the flow vector field. Although the highest priority is given to the visualization of the patterns themselves (or their absence), causes for their absence are also sought.

3 Choice of Methods

Here, we describe the criteria for our selection of visualization methods.

Visualization Goals

The main interest in the visualization of the in-cylinder flow is the extraction and visual analysis of the swirl and tumble motion patterns. Therefore, the flow vector field and its derived quantities are of primary interest.

For the use in design analysis, the constructed visualizations need to be objective and reproducible, meaning that the quality of the visualization result must not depend on vital parameters to be supplied by the user. This results in comparable visualizations for different simulation results of the same prototype or possibly even among different design prototypes. Therefore, in the selection of methods, we have put an emphasis on automatic schemes that require little or no user input.

Data Dimensionality

The simulation results are given in the form of attributes defined in the interior of the respective cylinder geometries. As is quite common in CFD simulations, the flow is required to vanish on the domain boundary (no-slip condition) in order to correctly model fluid-boundary friction. Nevertheless, values on the boundary of the domain are easily inferred by e.g. extrapolation of volume values next to the boundary. It is also notable that in classical engineering analysis, visualization is widely performed on two-dimensional slices.

Overall, the level of information that can be provided by a visualization technique increases with the dimension of the data it treats. At the same time, the visualization result need not necessarily improve due to perceptual issues such as cluttering. Finally, there is usually a price to pay in algorithmic complexity and computational cost as one progresses to higher dimensions. Therefore, for the case of our examples, we examine in some detail how the analysis of boundary and slice data allows to draw reliable conclusions on the pattern of the volume flow. We achieve this by a pairing of methods that combine boundary and volume techniques.

4 Extraction and Visualization of Swirl and Tumble Motion

In this section we present the methods along with corresponding visualization results and discuss their relevance with respect to the realization of the visualization goals. Due to limitation of available space for images, many of the figures used for illustration of individual methods actually show a combination of different visualization approaches. We discuss the benefits of such combinations in detail in Section 5. Moreover, the companion video [2] provides additional images and animations.

4.1 Global Flow Behavior using Integration-based Methods

Integration-based methods are well suited to the analysis of time-dependent flows. Their common application to stationary flows is only a special case. We study the applicability of this class of methods on two examples.

Particles and Pathlets

Despite their simplistic nature, particle visualization can provide valuable insight into the overall structure of a flow dataset (cf. e.g. [1, 11]). This is especially true for time-dependent data. While the basic principle is similar to that of streamlines or pathlines, an animation of moving massless particles manages to convey the dynamic nature of the flow much better than static imagery alone. In the general case, integral methods suffer from seeding issues, although strategies have been proposed to circumvent this (e.g. [18]). However, none of these approaches is concerned with time-varying data. Fortunately, engine geometries offer the inlet pipe as a natural choice of a seeding region. Integration of pathlines in time-dependent 3D flows is straightforward through the application of standard numerical integration algorithms that only require the integrand at a sparse set of points. While interpolation in time-varying grids is usually problematic, we were able to exploit the topologically invariant structure of the grid to simplify point location.

Figure 2 depicts a frame from an animation of massless particles moving with the flow during the early stage of the valve cycle, seeded at positions in the intake pipe. The particles are of uniform size and color-coded according to flow velocity magnitude. The image allows an easy identification of zones where the velocity is lower than average, hinting at a non-optimal inflow pattern at the side of the valve. Our general experience with this technique is that in spite of being visually imprecise, it greatly furthers the comprehension of the dynamic of the time-dependent flow by providing good overview.

Stream surfaces

Stream surfaces as an extension of streamlines are of great value in some applications because they manage to convey a spatially coherent picture of flow structures (cf. [3, 5, 13]). We found them to be of limited use in our case, for two reasons. First, both swirl and tumble are large-scale motions that are locally overlaid by other small-scale flow patterns and the resulting stream surfaces are complicated to interpret since the small details are emphasized by the surface nature of this primitive. Secondly, it is unclear how they can be applied satisfactorily in a time-dependent context. Although visualizations of some value can be generated, determination of good starting curves is strongly dependent on the dataset geometry. Hence, we do not consider stream surfaces an objective technique by the criterion stated above.

4.2 Topology-Based Visualization of Flow Structures

Topological methods provide efficient means for the visualization of essential structures in steady flows. As opposed to the integral methods described previously, they offer a fully automatic way to gain insight from vector data sets. The topological technique is typically applied in the visualization of planar flows [12] for which it yields synthetic graph representations. It consists of critical points (vector field zeros) and connecting separatrices. The three-dimensional case, however, remains challenging. Besides occlusion problems

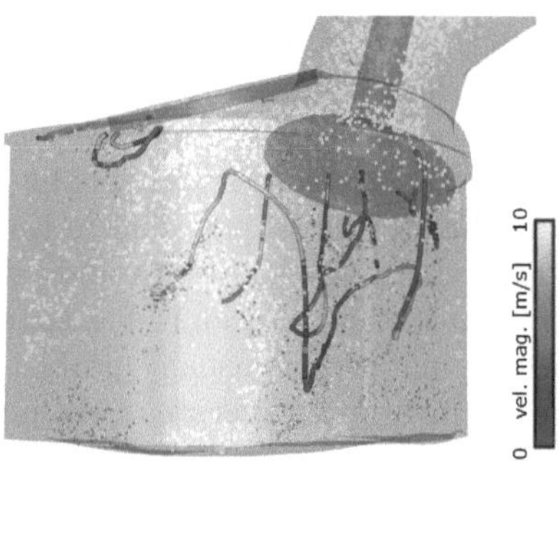

Fig. 2. A frame from a time-varying tumble motion visualization using a combination of particles and vortex cores lines. Particle velocity magnitude is color-coded. Lines are color coded by the path type (saddles – red, sources – green, sinks – blue). Over time, some of the particles are captured in the vicinity of vortex cores resulting in lost energy (lower velocity) for the creation of the tumble pattern (colorplate on p. 214).

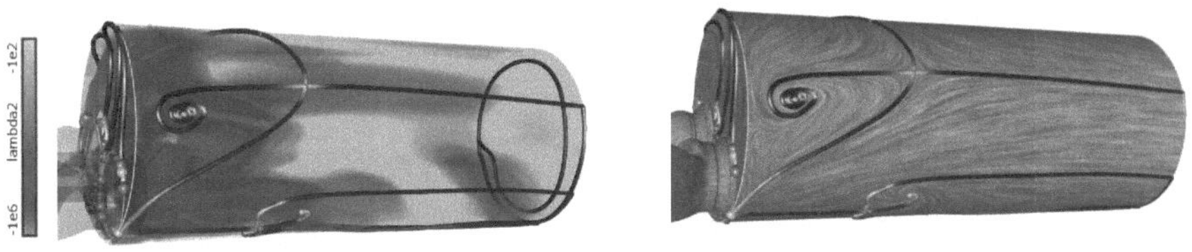

Fig. 3. Visualization of swirl motion using boundary topology. Critical points are colored by type (cf. Fig. 2), and separatrix color varies with separation/attachment behavior from dark blue (weak) to cyan (strong). Separatrices indicate the separation between neighboring vortices on the boundary. (Left) Combination with volume rendering with a transfer function of λ_2 only. On the bottom left of the cylinder, the recirculation zone causes a non-ideal off-center rotation, as visualized by topology. (Right) In combination with LIC (colorplate on p. 214).

that must be addressed specifically [15], prominent features of interest like vortices cannot generally be identified as elements of 3D local topology.

In this paper, we address these deficiencies by combining a topological analysis of the boundary flow with a hybrid approach that leverages 2D topology to explore the 3D structure of a vector field [16].

Boundary Topology

To our knowledge, a general algorithm for vector field topology on 2D unstructured grids embedded in three-space has not yet been described in the visualization literature (although algorithms exist for parameterized grids, e.g. [4]). Thus, we propose the following approach. In each triangle, we use the well-defined local tangent plane to perform a cell-wise search for critical points and determine their type. The construction of separatrices from saddle points is performed using a streamline integration approach based on geodesics as introduced by Polthier and Schmies [10]. A specific characteristic of triangulated surfaces for topological analysis is the existence of what we term *singular edges* [5]. Since the tangent plane is discontinuous across surface edges, the flow on both sides can be contradictory. Singular edges are typically found along the sharp contours of the geometry where they must be integrated in the topological analysis to account for the possibly contradictory flow behavior between neighboring cells.

For viscous flows, the information conveyed by the boundary topology can be enhanced naturally by showing the strength of flow separation and attachment along separatrices. Flow separation occurs when the flow surrounding an embedded body interrupts its tangential motion along the object's boundary and abruptly moves away from it. The opposite phenomenon is called

[5] Integral curves cannot continue over these edges since the interpolants on both sides of the edge are incompatible.

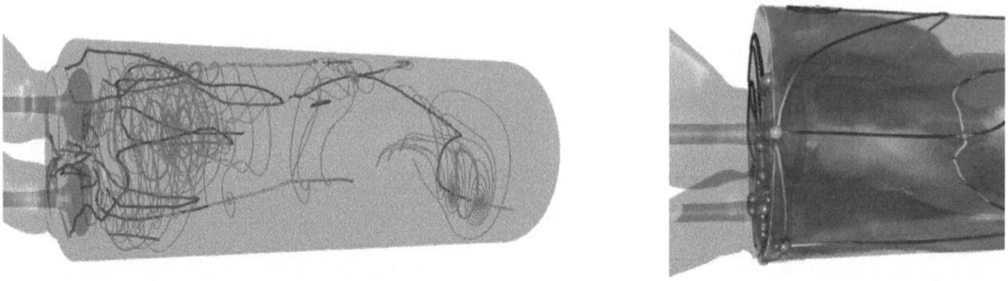

Fig. 4. *Left:* Cutting-plane topology applied to the diesel engine. Plane separatrices are colored gray. Despite the visual inexactness, swirl structures emerge clearly. Critical points paths are colored according to nature. It appears the overall swirl motion is fueled by several parallel vortices at the top of the cylinder. The main swirl motion core is disrupted near the middle. *Right:* Visualization of the rotational directions in the vortex system at the top of the diesel engine cylinder. The transfer function is identical to that of Fig. 6. The counter-rotating vortices appear in blue and in red depending on rotation direction. On the boundary, the topological analysis extracts and visualizes separation lines between individual vortices (color coding as in Figs. 2 and 3). (Colorplate on p. 214.)

flow attachment. As pointed out by Kenwright [7], separatrices of the boundary vector field constitute so called closed separation or attachment lines. To quantify flow separation and attachment along a separatrix, we compute the divergent (resp. convergent) behavior of neighboring streamlines by evaluating the local divergence of the vector field [17]. This is illustrated in Fig. 3 (left) by the color coding of the intensity of flow separation and attachment along separatrices.

Direct visualization of the boundary topology produces images such as Figure 3 (swirl dataset) and can also be applied in a time-dependent context. The combination of linear interpolation in time and in space often produces artifacts (such as artificial pairs of critical points that appear and quickly vanish quickly). Nevertheless, animations that show the temporal evolution of these instantaneous graphs provide valuable means to track the dynamics of important flow patterns (cf. Fig. 7). In our experience, boundary topology is most effective if used in conjunction with methods that visualize additional properties of the flow, both in the volume and on the boundary. Such combinations permit to determine the mutual influence between boundary and volume. We will discuss these issues when we examine further combinations of methods presented in Section 5.

Cutting-Plane Topology

It was shown previously [16] that a moving cutting plane that traverse the dataset and on which the vector field is resampled and projected at regular intervals can be a powerful tool in the analysis of 3D datasets. The projection of the vector field on the plane effectively manages to discard structures orthogo-

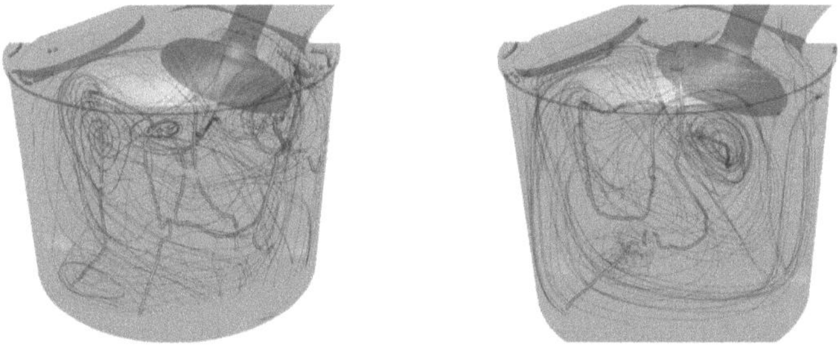

Fig. 5. Two frames from an animation of the tumble motion simulation. Cutting plane topology is applied to visualize flow field structures in the plane orthogonal to the tumble axis. Color of separatrices varies from blue to red on successive cutting planes. Tumble-like flow structures emerge clearly from the otherwise incoherent lines. The paths of critical points over the cutting plane continuum are displayed in green. In the last frame (right), the diagonal main tumble axis can be observed together with a large recirculation zone (closed path on the left). (Colorplate on p. 215.)

nal to the plane, but preserves plane-parallel flow patterns. If assumptions on the orientation of features are given, this property can be exploited. Cutting-planes are hence well suited for the qualitative analysis of swirl or tumble motion, since its axis of rotation is known. Furthermore, the (discretized) continuum of cutting-planes allows for the application of critical-point tracking over the plane parameter range. In the case of vortical motion that is intersected orthogonal to the rotational center, a reproduction of the vortex core as the path of a critical point over the parameter range should, in theory, work well. In practice, it is hard to intersect (not known a-priori) structures exactly. This results in approximations of the vortex cores. Still, a qualitative analysis is viable.

In Figure 4, the results of this approach applied to the swirl motion dataset are displayed. As the cutting planes are applied orthogonal to the cylinder axis, coherent swirl-type structures emerge at the top of the cylinder. Rotation cores orthogonal to the planes are visualized by critical point paths over the plane continuum. Interestingly, the main swirl core is supposed to extend through the whole cylinder, but is actually interrupted near the middle.

Figure 5 shows frames from an animation of the tumble dataset. The moving cutting planes have been applied orthogonal to the tumble axis and are color coded by their distance to the back wall of the combustion chamber for increased visual clarity. Although the visualization is not exact, the prevalent tumble structure is captured well in spite of its overall weakness and instability. Again, the center of the respective motions is given by the critical points paths. The tumble motion is found to consist of several smaller vortices, of which some have a diagonal orientation that looks like a simultaneous combination of swirl and tumble. In the full animation (cf. [2]), the interaction of the different smaller tumble patterns can be observed as they split and merge.

Using only the critical point paths for visualization, it is possible to observe the main tumble vortex, as shown in Figure 2. It does not completely match the desired axis and is highly off-center on the curved wall of the combustion chamber. Here too, the diagonal nature of the main rotation is confirmed.

Volume Visualization

Among the many region-based vortex definitions, the λ_2-criterion [6] has a well established tradition in engineering use. The criterion is given as a scalar quantity derived from the flow field Jacobian matrix and related to a minimum in pressure in the Navier-Stokes equation. According to its definition, a vortex is present at a point if $\lambda_2 < 0$. Traditionally, isosurfaces are used to visualize vortices. However, this approach does not fare well in complicated datasets with many vortices. The resulting isosurfaces do not separate individual vortices and are prone to visual complexity. This is only avoided by a careful manual selection of the isovalue. In addition, the strength of the rotation (as given by the modulus of λ_2) is not visualized. Interactive volume rendering has also been applied in this context [14, 16] and is able to overcome most of the difficulties related to isosurfaces. As described in [16], we have not attempted to apply volume rendering directly to the unstructured grid but have employed a resampling scheme that results in a rectilinear grid covering the region of interest (in our cases, the cylinder). This grid is then used in a direct volume rendering approach. Although artifacts are incurred in naive sampling, post-sampling scale-space filtering is very effective in removing these artifacts.

Using this approach, Figure 3 (left) illustrates the vortices in the context of the swirl motion using a simple one-dimensional transfer function that indicates vortex strength by color. As already visible in the results of the Sujudi-Haimes method, the actual swirl motion consists of several independent vortices, most prominently a strong vortex that spans almost the entire length of the cylinder. It is clearly off-center. It is apparent that the fuel-air mixture entering the cylinder is drawn into a strong rotational motion. Whether this is good for the overall mixing process or an obstacle that results in a pressure loss with negative consequences remains unclear. Technically, we have limited the transfer function range to λ_2-values between -10^6 and -10^2 to filter vortical motions at the small scales.

Recently, the use of multi-dimensional transfer functions [8] in flow visualization was investigated [16]. We make use of this technique to visualize the rotational orientation of individual vortices, allowing more insight into the mutual interaction in the observed vortex systems. Adding *normalized helicity* as a second variable, we manage to both strengthen the vortex region criterion as well determine the orientation of the rotational motion. Normalized helicity is given as the angle between the velocity vector and the vorticity vector and has a range of $[-1, 1]$, with positive values in the case of counter-clockwise rotation in flow direction and near-zero values indicating very weak vortical behavior. The obtained visualization (cf. Fig 4) indicates several mutually counter-rotating vortices in close proximity near the intake ports. This is highly undesired, since part of the energy contained in the inflow is used up by this vortex system.

The use of multidimensional transfer functions (used commonly in medical image generation) can require some degree of interactivity in the determination of the transfer functions in order to specify variable ranges manually for most satisfactory results. We wish to emphasize that in our case, transfer functions based on physical criteria (such as $\lambda_2 < 0$) are automatic in the sense the interesting value ranges are dictated by the laws of fluid dynamics. Using simple linear interpolation in time, an animation (cf. [2]) can depict the temporal evolution of vortices associated with swirl and tumble motion (see Fig. 6).

5 Hybrid approaches

While performing experiments with the different approaches detailed above, it became apparent that a combination of visualizations can provide an even more thorough understanding of the simulation results. In this section, we describe examples of particularly effective combinations and how they contribute to the swirl and tumble analysis.

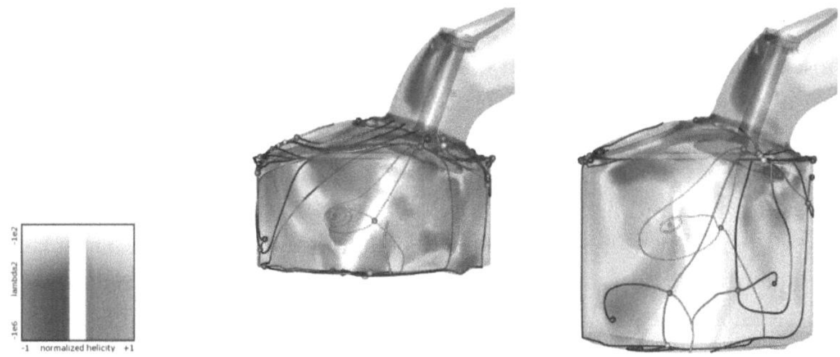

Fig. 6. Unsteady visualization of vortices from in-cylinder tumble motion in a gas engine and its relationship to the boundary. During the valve cycle (top left to bottom right), the piston head that shapes the bottom of the geometry moves down (not shown). The volume rendering shows vortices using a two-dimensional transfer function of λ_2 and normalized helicity (legend). The main tumble vortex is extracted and visible as off-center and with an undesired diagonal orientation. The flow structure on the boundary is visualized using boundary topology, with critical points colored according to their nature (cf. Fig. 2). A direct correspondence between the volume and boundary visualizations can be observed. In the third image, the intersection of the main vortex with the boundary results in critical points on the front and back walls.

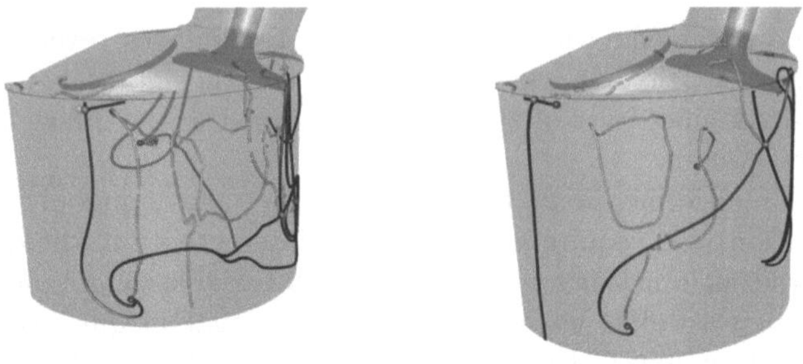

Fig. 7. Two frames from a time-varying visualization of tumble motion using a combination of cutting-plane topology (green lines) and boundary topology (blue lines). Where the critical point paths computed over the cutting-plane parameter range intersect the boundary, singularities appear there, too. This is a prime example of a hybrid approach being used to investigate the relationship between boundary and volume methods (colorplate on p. 215).

5.1 Boundary and cutting plane topology

In fluid flows, complex flow structures such as vortices are often caused by the interaction of the flow with boundaries. Depending on this interaction, different patterns appear on a boundary that in turn allow one to infer properties of the volume flow. However, in complicated geometries such as in-cylinder flow, this approach alone introduces visualization complexity due to the high number of vortices involved and generally complex flow structure.

The combination of boundary topology and cutting plane topology is an effective approach. Figure 7 illustrates this in time-slices from an animation. For the tumble flow, this type of visualization provides valuable insight into the development of the diagonal tumble motion. In the early stages of the valve cycle (left two images), the flow pattern is very incoherent and unstable. Roughly at the middle of the cycle (second image from the right), a swirl pattern occurs in the front half of the cylinder, mainly constituted by two large vortices. In the very last timesteps, the rotational axis tilts towards the desired tumble axis, but fails to reach it completely. Here, a large recirculation zone can be observed that may hinder the development of the tumble motion.

5.2 Sparse and dense methods

On the boundary, the topological graph as a visually sparse method is effectively combined with dense methods, such as texture-based methods or volume visualization. While texture-based methods are built on the capability of the human visual system to identify patterns in the flow, the topological graph serves as a terse structural picture that relies on cognitive interpolation on behalf of the viewer. It is therefore a very natural combination. Figure 3 (swirl, right image) provides an example, showing a very strong vortex near the inlet that is drawing away energy from the creation of the ideal swirl pattern. We refer the reader to previous work [9] for other applications of texture-based techniques in this problem context that we believe will benefit strongly from a pairing with a feature-based visualization.

A similar conclusion is reached by combining boundary topology and the volume visualization of vortices. Again looking for the swirl pattern, the imperfection of the actual flow motion is visible in Figure 3 (left). While the volume rendering shows the correctly oriented but off-center main vortex, the topology graph on the lower cylinder boundary complements this visualization by showing corner regions of the flow that are not taking part in the swirl pattern. As expected, the topological graph also serves to show separation and attachment boundaries that delimit the regions of influence of the different vortices. As confirmed by the rotational direction analysis (cf. Fig. 4), these vortices are rotating in different directions, which is considered destructive flow behavior.

There are of course other possible combinations: for example, the Sujudi-Haimes vortex core line visualization can play a similar role as cutting-plane topology. In our experiments we found however that it often detects only the strong vortices at the intake portions of both datasets. Combining particles and vortex core lines offers insights into how exactly the vortices are created and where kinetic energy is lost in small scale structures (cf. Fig. 2).

6 Conclusion and Future Work

By using a number of visualization techniques that we selected as automatic and objective, we were able to extract and create visualizations of the swirl and tumble datasets that allow an in-depth visual analysis of the actually occurring patterns. The visualizations are comparable between similar datasets and can thus be employed in design prototype analysis. By using hybrid combinations of different techniques, we were able to determine the extent to which the desired pattern is established and also detect influences that hinder its formation. In summary:

- The swirl motion in the diesel engine is visible in the form a prominent main vortex spanning the entire cylinder (Figure 3, left). It is non-optimally off-center. A recirculation zone is present in the lower corner of the cylinder. This may be the cause for the eccentricity of the swirl motion. A vortex system at the intake valve (Figures 3 and 4) prohibits the full conversion of energy contained in the incoming flow into the swirl motion.
- The observed tumble motion in the gas engine differs from an optimal pattern in several aspects:
  - it is unstable and sporadic over time
  - rather weak and off-center (Figs. 6 and 7) and
  - resembles a hybrid of both swirl and tumble patterns (Figures 2 and 7).

Here too, a prominent recirculation zone induces the swirl component into the flow. Again, the flow distribution at the intake valve is not optimal.

The unstructured, adaptive and time-varying nature of the tumble datasets poses a technical difficulty that we were able to circumnavigate through the choice of schemes and appropriate extensions where needed. The resulting visualizations are of high quality and provide valuable insight into the application.

There are many possible avenues that future work might take. High up on the priority list are improvements in the field of topological visualization. Full three-dimensional topology has not been completely realized, and while delivering viable visualization results, replacements such as cutting-plane topology are not completely satisfactory. The inclusion of features such as vortex cores into the topological skeleton is desirable, but may not be possible. Furthermore, the interplay between the topologies of the boundary and volume flows need to be investigated on a more systematic basis. In general, it seems desirable to examine the hybridization of different visualization approaches in the same context.

Future work could also feature improvements to many of the schemes presented here to allow for the treatments of larger datasets, such as entire engine blocks. It remains to be seen in how far this is feasible from a technical point of view.

Acknowledgements

The authors would like to thank all those who have contributed to this research including AVL (`www.avl.com`), the Austrian research program Kplus (`www.kplus.at`). The CFD simulation datasets are courtesy of AVL.

References

1. D. Bauer and R. Peikert. Vortex Tracking in Scale-Space. In *Data Visualization 2002. Proc. VisSym '02*, 2002.
2. C. Garth, R. S. Laramee, X. Tricoche, J. Schneider, and H. Hagen. Extraction and visualization of swirl and tumble motion from engine simulation data (companion video). `http://www.vrvis.at/scivis/laramee/MotionExtracted/`
3. C. Garth, X. Tricoche, T. Salzbrunn, and G. Scheuermann. Surface Techniques for Vortex Visualization. In *Proceedings Eurographics - IEEE TCVG Symposium on Visualization*, 2004.
4. A. Globus, C. Levit, and T. Lasinski. A tool for visualizing the topology if three-dimensional vector fields. In *IEEE Visualization Proceedings*, pages 33 – 40, October 1991.
5. J. P. M. Hultquist. Constructing Stream Surfaces in Steady 3D Vector Fields. In A. E. Kaufman and G. M. Nielson, editors, *IEEE Visualization '92*, pages 171 – 178, Boston, MA, 1992.

6. Jinhee Jeong and Fazle Hussain. On the Identification of a Vortex. *Journal of Fluid Mechanics*, pages 69–94, 285 1995.
7. D. Kenwright, C. Henze, and C. Levit. Feature extraction of separation and attachment lines. *IEEE Transactions on Visualization and Computer Graphics*, 5(2):135–144, 1994.
8. J. Kniss, G. Kindlmann, and C. Hansen. Multidimensional transfer functions for interactive volume rendering. *IEEE Transactions on Visualization and Computer Graphics*, 8(3):270–285, July-September 2002.
9. R. S. Laramee, D. Weiskopf, J. Schneider, and H. Hauser. Investigating Swirl and Tumble Flow with a Comparison of Visualization Techniques. In *Proceedings IEEE Visualization '04*, pages 51–58, 2004.
10. K. Polthier and M. Schmies. Straightest Geodesics on Polyhedral Surfaces. In Hans-Christian Hege and Konrad Polthier, editors, *Mathematical Visualization*, pages 135–150. Springer Verlag, Heidelberg, 1998.
11. F. H. Post, B. Vrolijk, H. Hauser, R. S. Laramee, and H. Doleisch. Feature Extraction and Visualization of Flow Fields. In *Eurographics 2002 State-of-the-Art Reports*, pages 69–100, 2–6 September 2002.
12. G. Scheuermann and X. Tricoche. Topological methods for flow visualization. In C.D. Hansen and C.R. Johnson, editors, *The Visualization Handbook*, pages 341–356. Elsevier, 2005.
13. D. Stalling. *Fast Texture-based Algorithms for Vector Field Visualization.* PhD thesis, Freie Universität Berlin, 1998.
14. S. Stegmaier and T. Ertl. A Graphics Hardware-based Vortex Detection and Visualization System. In *Proceedings of IEEE Visualization '04*, pages 195–202, 2004.
15. H. Theisel, T. Weinkauf, H.-C. Hege, and H.-P. Seidel. Saddle Connectors - An Approach to Visualizing the Topological Skeleton of Complex 3D Vector Fields. In *IEEE Visualization '03*, 2003.
16. X. Tricoche, C. Garth, G. Kindlmann, E. Deines, G. Scheuermann, M. Rütten, and C. Hansen. Vizualization of Intricate Flow Structures for Vortex Breakdown Analysis. In *Proceedings of IEEE Visualization*, pages 187–194, October 2004.
17. X. Tricoche, C. Garth, and G. Scheuermann. Fast and robust extraction of separation line features. In *Proceedings of the Dagstuhl Scientific Visualization Seminar*, 2003. to appear.
18. V. Verma, D. Kao, and A. Pang. A Flow-guided Streamline Seeding Strategy. In *Proceedings of IEEE Visualization '00*, 2000.

Simulation Methods for Advanced Design Engineering

Markus Trenker, Wolfgang Payer, and Matthias Haigis

Austrian Research Centers - arsenal research
Giefinggasse 2, A-1210 Vienna
markus.trenker@arsenal.ac.at

Summary. Since the time numerical simulation tools have been introduced in design engineering the requirements on their capabilities have increased steadily. Computer resources and people's expectations on the simulation itself have reached a high level over the years. More powerful computers are used to process meshes with several million cells. 3-d time accurate calculations with moving or deforming meshes are manageable as well as large eddy and direct numerical simulations. Advanced visualization techniques are used to extract more information leading to a deeper understanding of complex flow phenomena. Using examples of ongoing and recently finished projects the use of flow simulation methods and visualization tools will be presented. The use of commercial tools for cfd and post processing and their adaptation for simulation of human comfort as well as for particle transport simulations will be presented and potentials for topology based visualization methods will be pointed out.

1 Introduction

The development of visualization tools and the progress in numerical simulation methods is in close connection with the available computer power. Todays methods include 3-dimensional, time accurate simulations on meshes with several million cells. The massive amount of data does not only demand for highly optimized visualization tools in terms of storage management but also enables the user to get a deeper insight into local flow phenomena using advanced visualization methods like topology based analysis. In the following two examples will be presented which have great potential for a topology based visualization approach. The first one is concerned with passenger's thermal comfort in enclosed regions such as a cabin of an aircraft or a train's compartment. Disregarded for many years the topic of passengers thermal comfort draws much attention lately. Health aspects are in close connection to thermal comfort analysis and have to be considered as well in the design process of modern mass transportation vehicles. The second example addresses particle transport and accretion. Focussing on snow particles the application presented is concerned with snow accretion on the undercarriage of an urban train. Accumulation of snow on essential devices such as air inlets and outlets of the engine cooling system which are mounted underfloor can lead to overheating and malfunction and finally to serious damage. This simulation method is also applicable to snow accretion on aircrafts and road vehicles and in civil engineering to simulate snow accumulation on buildings, bridges, avalanche protection devices and freeways. The flow simulations as well as the post processing are performed using commercial tools. Although these tools are state of the art their capabilities for topological flow analysis are very limited.

2 Thermal comfort of passengers

The human response to the thermal environment is depending on the six major factors air temperature and velocity, mean radiant temperature, relative humidity, physical activity and clothing thermal resistance, [5]. Comfort is by definition a subjective sensation. However, the use of standardized scales for ratings of thermal sensations for large groups of individuals has enabled researchers to define the essential components of the thermal environment and their interaction in determining the average response of the population. In practice the indoor climate is far from being uniform. Cold drafts from windows, supply of air from different ventilation systems etc. contribute to the creation of various comfort asymmetries. For engineering purposes the term "comfort" has to be converted and expressed in measurable, physical quantities, [13]. Commonly used parameters for evaluating passenger comfort are the predicted mean vote PMV, the predicted percentage dissatisfied PPD, the equivalence temperature t_{eq} and the mean age of air MAA.

Predicted mean vote PMV and the predicted percentage dissatisfied PPD

In order to get quantitative results for the definition of comfort Fanger used the predicted mean vote model. Using this method a steady state thermal comfort index is derived from the human heat balance calculations and climate chamber studies, (1). It relates the PMV of a group of people exposed to a certain environment to the calculated result of the human heat balance equation and is given by

$$PMV = \left(0.303e^{-0.036(M-W)} + 0.028\right)\left((M-W) - 3.05 \times 10^{-3} \times \right.$$
$$[5733 - 6.99\,(M-W) - p_a] - 0.42[(M-W) - 58.15] - 1.7 \times 10^{-5} M \times$$
$$(5867 - p_a) - 0.0014M\,(34 - t_a) - 3.96 \times 10^{-8} f_{cl}[(t_{cl} + 273)^4 -$$
$$\left. - \left(\overline{t_r} + 273\right)^4] - f_{cl} h_c\,(t_{cl} - t_a)\right)$$
$$\text{where} \quad (1)$$
$$t_{cl} = 35.7 - 0.028\,(M-W) - I_{cl}\Big[3.96 \times 10^{-8} f_{cl}\Big((t_{cl} + 273)^4$$
$$- \left(\overline{t_r} + 273\right)^4\Big) + f_{cl} h_c\,(t_{cl} - t_a)\Big]$$

with M the metabolic rate in W/m^2 and W the external work in W/m^2, t_a the ambient air temperature and $\overline{t_r}$ the mean radiant temperature of the environment in $°C$ and p_a the water vapor pressure in Pa, [4]. f_{cl} stands for the clothing area factor, which is 1 for unclothed situations and I_{cl} is the clothing insulation in clo. The surface temperature t_{cl} of the clothing in $°C$ has to be calculated iteratively. The PMV thermal sensation scale is used by probands to rate their feeling of comfort on a scale from -3, which corresponds to "too cold" to +3, which stands for "too hot", see fig. 4. The PPD is directly related to the PMV [4] by

$$PPD = 100 - 95e^{-n}$$
$$\text{where} \quad (2)$$
$$n = 0.03353PMV^4 + 0.2179PMV^2.$$

While PPD provides information as to whether the environment is likely to be acceptable, PMV gives information whether it is too hot or too cold when the number dissatisfied is too large, [12].

Equivalence temperature t_{eq}

t_{eq} is defined as the temperature of an imaginary enclosure with the mean radiant temperature equal to air temperature and still air in which a person has the same heat exchange by convection and radiation as in the actual conditions, [13]. In cfd-calculations t_{eq} can be derived in various ways. Many empirically derived models are available in the literature, [2]. We are using a formulation suggested by Madsen et al. [9]

$$t_{eq} = 0.5\left(t_a + \overline{t_r}\right) \qquad \text{for} \qquad u \leq 0.1m/s$$
$$t_{eq} = 0.55t_a + 0.45\overline{t_r} + \frac{0.24 - 0.75\sqrt{u}}{1 + I_{cl}}\,(36.5 - t_a) \qquad \text{for} \qquad u > 0.1m/s \quad (3)$$

with a human energy metabolism of about 70 W/m^2 and u the local velocity in m/s. Figure 3 shows the equivalence temperature for the thermal mannequin as depicted in fig. 6. The mannequin is seated in a train compartment with air inlets on the bottom in front of the mannequin and at the window. The inlet velocity is 2 m/s at the static temperature of 291K. The radiation caused by the sun is 436 W/m^2.

Mean age of air *MAA*

The age of air is a useful parameter to evaluate artificial ventilation effectiveness. The age of air in a point is the time taken by air flowing out of the inlets to reach that point, [11], [10]. It is by definition zero at the inlets. For the scalar ϕ the transport equation (4) has to be solved in i dimensions

$$\frac{\partial \rho \phi}{\partial t} + \frac{\partial}{\partial x_i}\left(\rho u_i \phi - \Gamma \frac{\partial \phi}{\partial x_i}\right) = S_\phi \qquad \text{for } i = 1 \dots 3 \tag{4}$$

with ρ the density, t the time, x_i the cartesian coordinates and u_i the velocities in i directions. Γ and S_ϕ are the diffusion coefficient and the source term and ϕ the age of air.

2.1 Thermal comfort in modern train compartment

The train under investigation is a modern two floor rapid transit train, [7]. Simulation is performed for the complete internal air flow, which mainly consists of two parts as depicted in fig. 1. The ventilation system transports air from the air conditioning system to the inlet vents of the compartment.

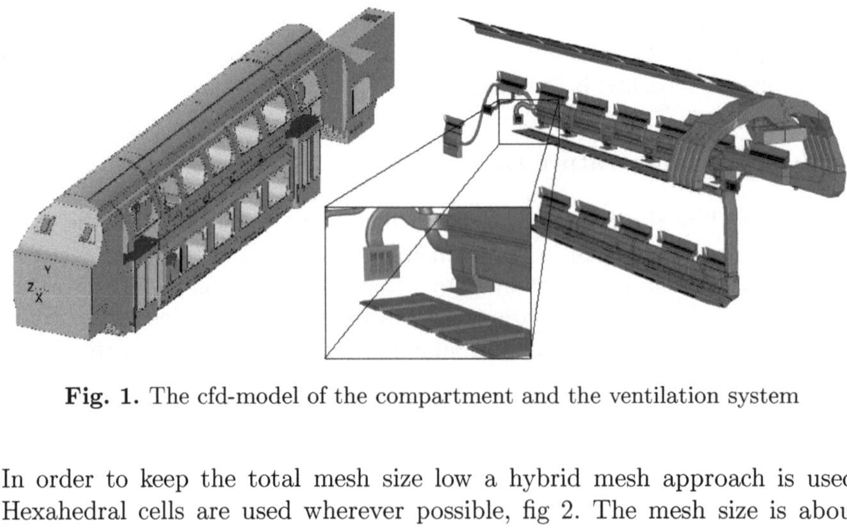

Fig. 1. The cfd-model of the compartment and the ventilation system

In order to keep the total mesh size low a hybrid mesh approach is used. Hexahedral cells are used wherever possible, fig 2. The mesh size is about

15 millon cells. More detailed information about the flow simulation can be found in [7].

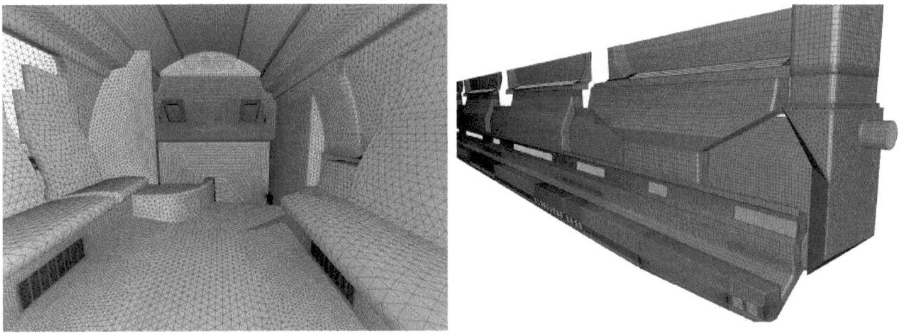

Fig. 2. The surface meshes inside the train's upper floor and on the floor and window inlets of the ventilation system

Figure 3 shows the equivalence temperature for the thermal mannequin in fig. 6 with local I_{cl} values set for a normally dressed man. t_{eq} is calculated for each segment, such as head, chest, arms etc.. Shaded regions mark comfort regions assessed from ratings of a group of people exposed to a certain environment, [13]. Having all t_{eq} values inside any of the "but comfortable" zones will not result in a general acceptance for the whole body. For a general acceptance a weighted whole body t_{eq} value has to be calculated and all local values have to be close to that value in any but the "too cold" or "too hot" regions. The comfort regions in fig. 3 are calculated for the summer months. As can be seen in fig. 3 the upper body segments are in the "hot but comfortable" region with a higher t_{eq} on the mannequins left side due to the radiation from the sun.

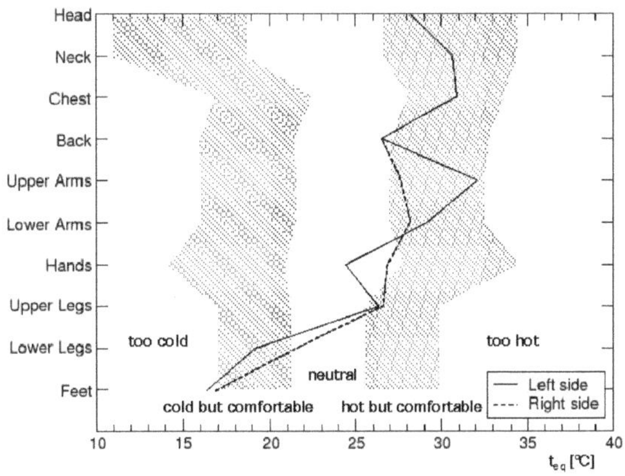

Fig. 3. t_{eq} with limitations for summer months of the thermal mannequin in fig. 6

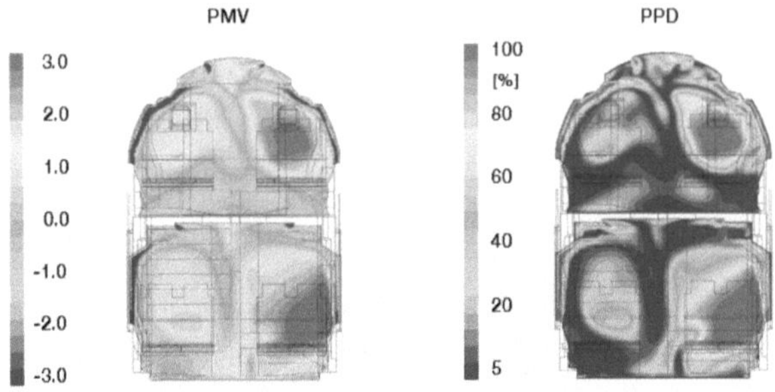

Fig. 4. *PMV* and *PPD* values for a section of the train, [7] (colorplate on p. 216).

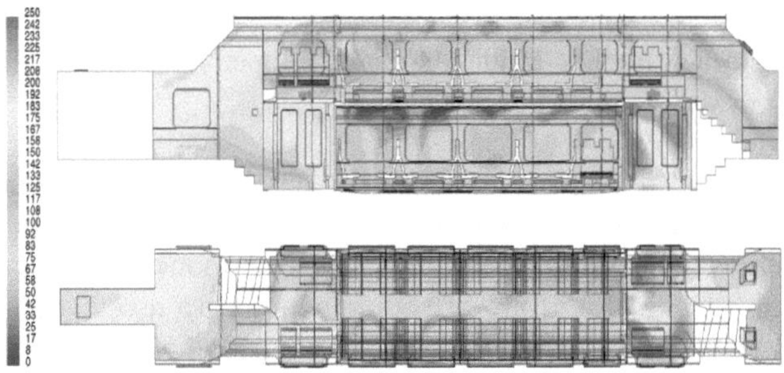

Fig. 5. *MAA* in [s] in the train, [7] (colorplate on p. 216).

Figure 4 shows the *PMV* in a section of the train. Since the sun is shining from the left side in the compartment high *PMV* values are found close to the windows on this side. Cold air is transported by the inlets into the compartment. Here the temperature is locally too cold which results in low *PMV* values. It is well known that different people will have a different perception of the climate produced and that any given climate is unlikely to be considered satisfactory by all. Therefore satisfying 80% of occupants is good, so that a *PPD* of less than 20% is good, [12]. For enclosed areas with occupants *MAA* is used as an indicator for local air quality. More detailed information on air quality can be assessed by calculating the species transport of air. However for large applications such as the complete internal flow of a compartment *MAA* gives sufficient information on air quality. As in this example it can be observed in fig. 5 that the oldest air is caught on the stairways and in the exit regions.

Topological flow analysis

The picture on the right in Figure 6 shows a topological flow analysis of the upper part of the body. Using commercial postprocessing tools [6] a LIC (Line Integral Convolution) algorithm is used to visualize wall stream lines. The

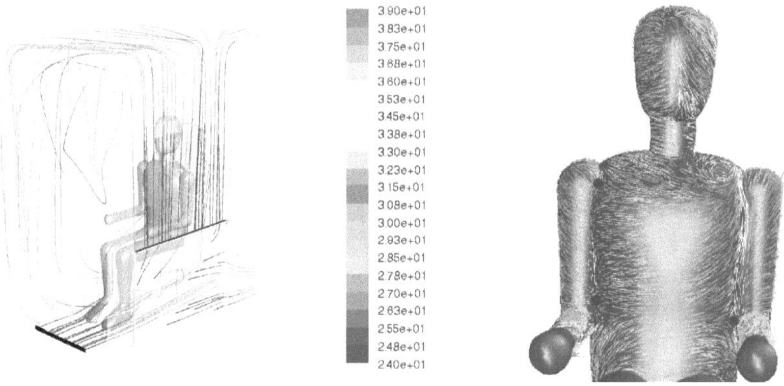

Fig. 6. The thermal mannequin in the compartment. Left: Visualization of streamlines, coloured with their residence time in [s]. Surface static temperature from 291K to 310K. Right: Texture based visualization of wall stream lines coloured with the local t_{eq} in $^{\circ}C$ (colorplate on p. 216).

colours correspond to the local t_{eq}. Critical points are found where the vector field magnitude vanishes. Following the classification of critical points found in [8] two attracting foci (red stars) and two saddle points (blue dots) are found on the front side of the upper body. Its interesting to note that the positions of critical points correspond to the location of regions with high values of t_{eq}. The visualization of wall stream lines gives detailed information on how the local t_{eq} is transported on the body surface. Together with the visualization of streamlines in the compartment and the static surface temperature of the dummy this information is extremely useful for optimizing thermal passenger comfort. Due to the high complexity of the geometry of the compartment and complex variables to be displayed a context based feature visualization as presented in [19] would be very useful. Areas of interest with detailed information on comfort related parameters are selectable where as the rest is displayed in a coarse manner. Cross cutting topology as presented in [18] is another promising method for analyzing flow topology. It allows to apply existing algorithms for topology tracking of two-dimensional vector fields to the visualization of three-dimensional flow structures. In this manner 2-d cut planes are defined in the continuous three-dimensional physical space of the original data. Detection of critical points and separatrices are carried out in these 2-d cut planes and can be analyzed with respect to comfort relevant parameters.

3 Particle transportation and accretion

Snow accretion plays an important role for the operation of vehicles in countries with frequent snowfall. In the case of a train running on snow covered

tracks, snow is being dispersed from the ground and accumulates on parts of the undercarriage. This can result in blockage of air inlets and outlets of the engine cooling system and various other essential devices, finally leading to vehicle failure. In one of the first publication about saltation Bagnold [1] identified three distinct transport modes for the transport of particles by an air stream: surface creep, saltation and suspension. For the accretion of snow on train undercarriage suspension is supposed to be the main reason.
To quantify the saltating mass flux depending on the local friction velocity a number of mostly empirical models can be found in the literature. White [20] who initially investigated soil transport by winds on Mars, found a correlation, borne out of microphysical saltation models as well as wind tunnel studies. The model also holds for conditions on earth and is given in (c.g.s.) units by

$$F = \frac{c_s \rho {u_\star}^3}{g}\left(1 - \frac{u_{\star tf}}{u_\star}\right)\left(1 + \frac{{u_{\star tf}}^2}{{u_\star}^2}\right) \tag{5}$$

with F the mass flux per unit width, $u_\star = \sqrt{\tau_w/\rho}$ the friction velocity, ρ the density of the fluid, τ_w the wall shear stress and $c_s = 2.61$. The fluid threshold friction velocity $u_{\star tf}$ has to be exceeded for suspension to occur.
Sorensen [15] modified Bagnold's formulation in [1] and derived the equation,

$$F = 0.0014 \rho u_\star \left(u_\star - u_{\star c}\right)\left(u_\star + 7.6 u_{\star c} + 205\right) \tag{6}$$

for the mass flux in (c.g.s) units, which is an approximately cubic function of $u_\star$. The critical friction velocity is set to $u_{\star tf}$ in the following. Although these models were developed for sand, they are in good agreement with experiments also for snow particles and are therefore widely used in the literature, e.g. [14, 3]. Figure 7 shows the mass flux per unit width versus the friction velocity for the two models under consideration.

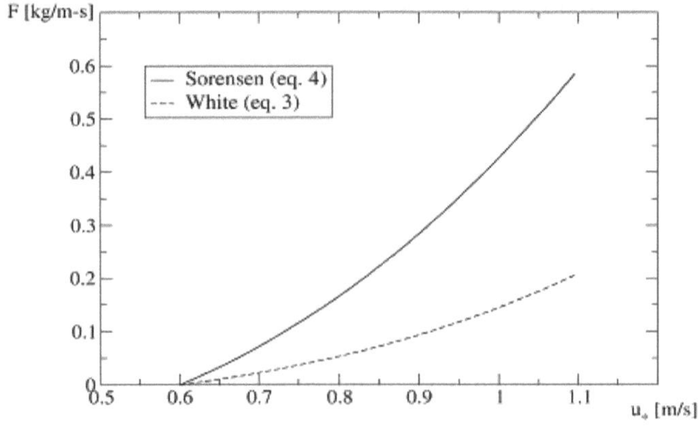

Fig. 7. The particle mass flux plotted versus the friction velocity

For quantifying snow saltation several differences in grain characteristics compared to sand particles are obvious. Newly fallen snow has a dendritic (greek. *dendron* tree) shape rather than a spherical shape which will influence the aerodynamic behaviour during suspension and saltation. Due to their distinct shape, snow particles will as well differ in their structure-elastic behaviour compared to sand particles. This is important for finding correct values for the threshold velocities. Snow particles being transported by an airstream will also change in shape due to collisions with neighbour particles and in that way alter the overall snow properties, [3]. The simplicity of the aforementioned models together with their use for a wide range of engineering applications concerning particle saltation found in the literature confirms the use of these models to simulate snow accretion. Due to the lack of any experimental data the model introduced by Sorensen is used. This model gives higher mass fluxes compared to the model by White and is therefore the conservative choice for the calculation of accretion on critical surfaces.

3.1 Snow accretion on a train undercarriage

The train under investigation has a total length of 54 meters. Its cad-model, already simplified for cfd, is depicted in fig. 8. In the simulation, the tracks and the ground are moving with the inlet velocity of 22.2 m/s, the wheels and axels of the train are kept non-rotating, because it has only marginal effects for the given problem. The flow field is simulated by solving the steady Reynolds

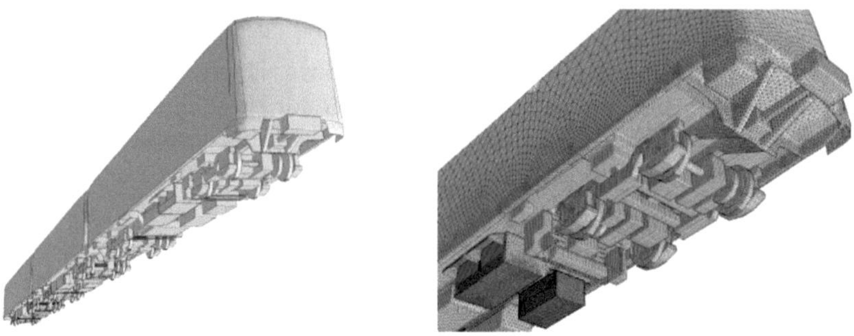

Fig. 8. The cfd-model of the train with a detailed view of the surface mesh

averaged Navier-Stokes equations employing the standard k-ω SST turbulence model with standard wall functions, [6]. In addition to solving the transport equations for the continuous phase, a discrete second phase is introduced in a Lagrangian frame of reference, following the Euler-Lagrange approach. This second phase consists of spherical particles dispersed in the continuous phase. Trajectories of these discrete phase entities are computed. A fundamental assumption made in this model is that the dispersed second phase occupies a low volume fraction, even though high mass loading is acceptable. It is assumed, that the snow concentration in the air is small enough to have no significant influence on the solution of the flow field. The particle trajectories are computed individually after the fluid phase calculation has reached the convergence criterion. The particle mass flux is calculated using (6) with $u_{\star tf} = 0.21$ m/s, according to [14]. More detailed information about the flow simulation can be found in [17]. As particles are being transported by the flow, they are either trapped or reflected on walls. The trapping of particles depends on the local fluid threshold friction velocity. It is assumed that particles hitting the surface with a local friction velocity smaller than $u_{\star tf} = 0.21$ m/s will be trapped. For particles being reflected on a wall the momentum of the particle after the impact is altered to ensure that reflected particles travel alongside the wall instead of an elastic impact. For an average snow particle the diameter was set to 0.48 mm according to [14]. The density of snow particle is set to 366 kg/m^3.

Figure 9 shows the accretion of snow on parts of the undercarriage. These parts of surfaces which are exposed to the main flow show higher accretion values than surfaces hidden or protected by others. Almost no accretion is found on the outlet, since the averaged velocity of 4.85 m/s prevents particles from being adhered. The compressor inlet is shown in fig. 10. This inlet is located on the upper side of the compressor box and is assumable well hidden from the main flow. This inlet shows higher accretion rates compared to the air inlet in fig. 9, which is much more exposed to the main flow. The rather narrow gap between the compressor inlet and the underfloor of the train leads to local pressures below the ambient pressure, which is the main reason for particles being sucked into the compressor inlet. Figure 10 reveals that the location of the compressor's mounting strut above the inlet has great influence on the local flow field and is partly responsible for these high accretion rates. The particles tracks are post-processed in reverse. This ensures that the paths of all particles which will finally stick on the compressor inlet are tracked from their beginning.

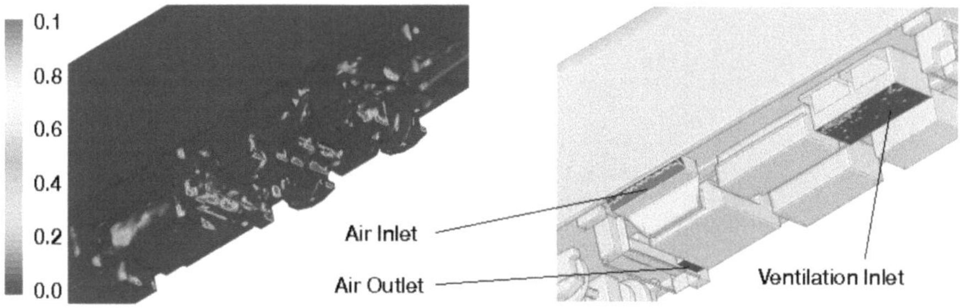

Fig. 9. Snow accretion in kg/m^2s on wall boundaries (colorplate on p. 217).

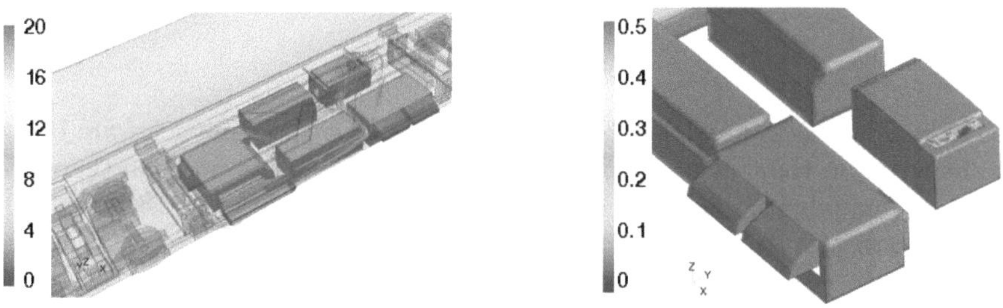

Fig. 10. Snow accretion in kg/m^2s on the compressor inlet and particle tracks coloured with the local particle velocity in m/s collected by the compressor (colorplate on p. 217).

Topological flow analysis

For studying particle accretion on a surface the flow around a simple box is investigated in free stream conditions similar to the train's flow conditions as depicted in fig.11. Particles are injected at the far field inlet and are transported by the free stream. Calculations were performed for different angles of attack of the box's top side. The box itself is closed for all angles of attack of the top side. Pictures p1 and p2 show wall stream lines using a LIC algorithm coloured with the local wall shear stress. Increasing the angle of attack moves the saddle point (blue dot) upstream and in that way shortens the recirculation area. Picture p3 shows the accretion rate on that surface. Since particles stick to the surface only in areas where $u_\star$ and therefore τ_w is smaller than their threshold values the accretion rate is high close to the separatrice (orange line) connecting the 2 foci (red stars) and the saddle point. In [16] an approach was made to study flow topology based stream lines and path lines. This approach is of particular interest since snow particles can approximately

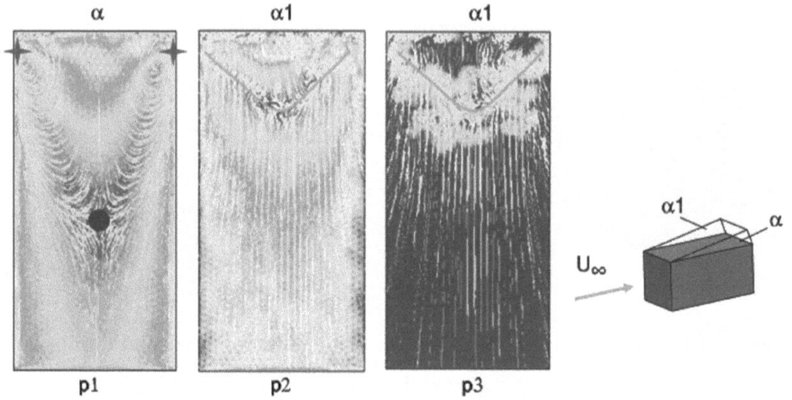

Fig. 11. Visualization of wall stream lines and accretion rate together with critical points for different angles of attack of the top surface of the box (colorplate on p. 217).

be seen as massless particles. Under this assumption snow particles are being transported in the flow along path lines. This proposed method can therefore be used to analyze local attractors on surfaces exposed to particle flow which enforce particles to accrete.

4 Conclusions

The two examples presented show the complex demands on todays numerical simulation methods. Changing requirements from the industry as well as a deeper understanding of flow phenomenons lead to new approaches for visualization. The work presented in this contribution shows the importance of topology based analysis but also the limited capabilities of post processing tools available today. As for the passengers thermal comfort human physiology has to be included into traditional flow simulation. The interaction between human physiology and the surrounding local climate is still not well understood. Using topology based approaches the analysis of critical points and separatrices on thermal mannequins would give new insights in passengers thermal comfort. In unsteady flows structural changes of the flow field, i.e. bifurcations in the vicinity of the mannequin might be used to analyze the threshold between human comfort and discomfort. For combined simulations of continuous and discrete phases finding and categorize critical points on walls is crucial for analyzing particle accretion. To understand those flows improved methods for visualizing must be available making detailed flow information transparent to the engineer. A cooperation of all the disciplines involved is necessary for a efficient development of improved tools for flow simulations and visualization.

References

1. Bagnold, R.A.: The Physics of Blown Sand and Desert Dunes. Methuen, London (1941)
2. Bedford, T.: The warmth factor in comfort at work. MRC Industrial Health Board Report HMSO, London, UK, **76** (1936)
3. Doorschot, J.J.J., Lehning, M.: Equilibrium Saltation: Mass fluxes, Aerodynamic Entrainment and Dependence on Grain Properties. Boundary-Layer Meteorology, Kluwer, **104(1)**, 111–130, (2002)
4. Ermittlung des PMV und des PPD und Beschreibung der Bedingungen fr thermische Behaglichkeit. EN ISO 7730 (1995)
5. Fanger, P.O.: Thermal Comfort: Analysis and Applications in Environmental Engineering. Danish Technical Press, Copenhagen (1970)
6. Fluent 6.1.22, 2004, User Manual
7. Haigis, M., Haider, G., Payer, W., Mann, M.: Comfort Simulation of a Double Deck Train Coach using CFD Analysis. 9. International Conference on Air Distribution in Rooms, Coimbra, Portugal (2004)
8. Helman, J.L., Hesselink, L.: Visualizing Vector Field Topology in Fluid Flows. IEEE Computer Graphics and Applications, **11(3)**, 36-46 (1991)
9. Madsen, T., Olesen, B., Kristensen, N.: Comparison between operative and equivalent temperature under typical indoor conditions. ASHRAE Transactions, **90**, 1, 1077–1090 (1984)
10. Mantegna M.: An Invariant of Age of Air: Proof and Applications. Air Infiltration Review, **14**, 2, 1993
11. "A guide to air change efficiency", Tech Note AIVC, Coventry, UK, 1980
12. Memarzadeh, P.E., Manning, A.: Thermal Comfort, Uniformity and Ventilation Effectiveness in Patient Rooms: Performance Assessment Using Ventilation Indices. ASHRAE, Annual Meeting, **106**, 2, 748–761, 2000
13. Nilsson, H.O.: Comfort Climate Evaluation with thermal Manikin Methods and Computer Simulation Models. Thesis, Royal Institute of Technology, University of Sweden, Sweden (2004)
14. Nishimura, K., Hunt, J.C.R.: Saltation and incipient suspension above a flat particle bed below a turbulent boundary layer. J. Fluid Mech., **417**, 77–102, 2000
15. Sorensen, M.: An analytic model of wind-blown sand transport. Acta Mech. Suppl. 1, 67–82 (1991)
16. Theisel, H., Weinkauf, T., Hege, H.-C., Seidel, H.-P.: Topological Methods for 2D Time-Dependent Vector Fields Based on Stream Lines and Path Lines. IEEE Transactions on Visualization and Computer Graphics, **11**, 4, 383–394, 2005
17. Trenker, M., Payer, W., Krenn, C., Haider, G., Mann, M.: Numerical Simulation of Snow Entrainment with Application to Train Undercarriage Design, NAFEMS World Congress, Malta (2005)
18. Tricoche, X., Garth, C., Kindlmann, G., Deines, E., Scheuermann, G., Ruetten, M., Hansen, C.: Visualization of Intricate Flow Structures for Vortex Breakdown Analysis, to appear in Proceedings of IEEE Visualization '04, Austin TX, October 2004
19. Viola, I., Gröller, E.: Focus+Context Visualization of Features and Topological Structures, presented at the TopoInVis, Budmerice, September 2005
20. White, B.R.: Soil Transport by winds on Mars. J. Geophys. Res., **84**, 4643–4651 (1979)

A Practical Approach to Two-Dimensional Scalar Topology

Peer-Timo Bremer and Valerio Pascucci

Center for Applied Scientific Computing
Lawrence Livermore National Laboratory
{pascucci1,ptbremer}@llnl.gov

Summary. Computing and analyzing the topology of scalar fields has proven to be a powerful tool in a wide variety of applications. In recent years the field has evolved from computing contour trees of two-dimensional functions to Reeb graphs of general two-manifolds, analyzing the topology of time-dependent volumes, and finally to creating Morse-Smale complexes of two and three dimensional functions. However, apart from theoretical advances practical applications depend on the development of robust and easy to implement algorithms. The progression from initial to practical algorithms is evident, for example, in the contour tree computation where the latest algorithms consist of no more than a couple of dozens lines of pseudo-code. In this paper we describe a similarly simple approach to compute progressive Morse-Smale complexes of functions over two-manifolds. We discuss compact and transparent data-structures used to compute and store Morse-Smale complexes and demonstrate how they can be used to implement interactive topology based simplification. In particular, we show how special cases arising, for example, from manifolds with boundaries or highly quantized functions are handled effectively. Overall the new algorithm is easier to implement and more efficient both run-time and storage wise than previous approaches by avoiding to refine a given triangulation.

1 Introduction

Scalar field topology has proven to be a powerful tool in a wide variety of applications. In scientific visualization it has been used, for example, for speeding up the extraction of iso-surfaces [25], simplifying them [4], and also for general data analysis [3, 11]. However, topology is also useful as a shape descriptor [22, 14], can help in remeshing surfaces [24, 7], and widely used concepts like watershed and flood-fill algorithms [12] are also topological in nature.

The most complete description of the topology of a scalar field is its *Morse-Smale complex* which segments the field based on its gradient. While the usefulness of the Morse-Smale complex has been widely acknowledged it has rarely been applied in practice. The problem is that intuitive straightforward computations of the Morse-Smale complex which have been described initially [23, 2] do not work in general. Only recently, generically correct algorithms to compute two- and three-dimensional Morse-Smale complexes have been discovered [9, 8]. However, these algorithms are notoriously difficult to implement which again limits their use in practice. Only the third generation of algorithms [3, 11] approach usability by combining correctness with a reasonable implementation.

In this paper we discuss a further improvement on the algorithms for computing the Morse-Smale complex of two-dimensional functions. In particular, we want to encourage the use of topology in practice. Therefore, the paper focuses not on any particular scientific result using Morse-Smale complexes but rather on providing a simple but complete description of the data structures and algorithms necessary to apply the theory of Morse-Smale complexes in general.

1.1 Related Work

Under various names the Morse-Smale complex has been used in a number of different contexts. Initially, they were used to describe the topology of terrains [5, 17] or other functions that could be connected to geography [26]. Naturally, the early research was limited to theoretical definitions rather than computer based algorithms. This line of research also includes the definitions of early multi-resolution structures [20, 21] as well as an in-depth discussion about potential degeneracies arising from non-smooth functions [19].

The first straight forward implementations are described by Takahashi et al. [23] and Bajaj and Schikore [1]. However, neither paper discusses any of the degeneracies (e.g. areas of zero gradient, merging integral lines) which arise when applying smooth theory to piece-wise linear functions. Therefore, these algorithms are likely to run into difficulties on any scalar field with non-trivial topology. The algorithms discussed in this paper are most closely related to the ones by Edelsbrunner et al. [9] and Bremer et al. [3]. Edelsbrunner et al. describe extensively how to handle merging and splitting steepest paths. However, their algorithm is quite involved and its implementation error prone. Bremer et al. simplify the implementation significantly by showing that only certain steepest lines must be kept separate and suggest to refine the mesh in order to avoid the complicated data structures of Edelsbrunner et al. Unfortunately, this approach, by definition, requires a dynamic mesh data structure able to refine the mesh at run-time. This is a distinct disadvantage when dealing with large data sets since it prevents the use of highly compact but rigid mesh encodings. Here we use a further extension of the algorithm in [3]. In Sect. 3 we show that not only can certain steepest lines be merged but those that must be kept separated can only create a very limited number of local configurations. We show how to encode these configurations efficiently and thus mostly avoid refining the mesh. Furthermore, Sect. 4 describes how a Morse-Smale complex with boundary is simplified and how to locally determine all possible simplifications.

The remainder of the paper is organized as follows: Section 2 introduces all necessary concepts of Morse theory and provides a collection of terms used throughout the paper. Section 3 discusses how to compute a Morse-Smale complex of functions on triangulated two-manifolds. Finally, Sect. 4 deals with simplifying a Morse-Smale complex with boundary and discusses a simple data structure to store a complex progressively.

2 Theory

In this section we introduce the necessary concepts from smooth and piecewise linear Morse theory. We refer to [18] and [16] for further background.

2.1 Morse-Smale Complexes

Throughout this paper, $\mathbb{M}$ denotes a compact 2-manifold without boundary and $f : \mathbb{M} \to \mathbb{R}$ denotes a real-valued smooth function on $\mathbb{M}$. Assuming a local coordinate system (x, y) at a point $a \in \mathbb{M}$, the point is called *critical* if its *gradient* $\nabla f(a) = (\partial f/\partial x, \partial f/\partial y)$ vanishes and called *regular* otherwise. Examples of critical points are maxima (f decreases in all directions), minima (f increases in all directions), and saddles (f switches between decreasing and increasing more than twice around the point).

Using the local coordinates at a, we compute the *Hessian* of f denoted by $H(a)$, which is the matrix of second order partial derivatives. A critical point is *non-degenerate* if the Hessian is non-singular, which is a property that is independent of the local coordinate system. According to the Morse Lemma, it is possible to construct a local coordinate system such that f has the form $f(x, y) = f(a) \pm x^2 \pm y^2$ in a neighborhood of a non-degenerate critical point a. The number of minus signs is the *index* of a and distinguishes the different types of critical points: minima have index 0, saddles have index 1, and maxima have index 2. The function f is a *Morse function* when all its critical points are non-degenerate and have pairwise different function values.

At any regular point, the gradient (vector) is non-zero, and when we follow the gradient we trace out an *integral line*, which starts at a critical point and ends at a critical point, while technically not containing either of them. Since f is smooth, two integral lines are either disjoint or the same. The *descending manifold* $D(a)$ of a critical point a is the set of points that flow toward a. More formally, it is the union of a and all integral lines that end at a. The collection of descending manifolds is a complex in the sense that the boundary of a cell is the union of lower-dimensional cells. For example, the descending manifolds of maxima are open discs, whose boundary consists of the descending manifolds of saddles (open intervals) which in turn are bounded by (descending manifolds of) minima (points). Symmetrically, we define the *ascending manifold* $A(a)$ of a as the union of a and all integral lines that start at a. If no integral line starts and ends at a saddle, see [9], we can overlay these two complexes and obtain the *Morse-Smale complex* (MS complex) of f. Its *nodes* are the vertices of the two overlayed complexes, which are the minima, maxima, and saddles of f. Its *arcs* are integral lines starting or ending at saddles, and its *regions* are areas bounded by four arcs. An example is shown in Fig. 8(a). The MS complex provides a complete topological segmentation of $\mathbb{M}$ and thus is the fundamental theoretical structure many popular algorithms like contour-tree extractions or watershed computations are build upon.

2.2 PL Morse Theory

Unfortunately, traditional Morse theory is very dependent on f being smooth; a requirement rarely fulfilled in practice. Usually, one deals with piece-wise linear (pl-) functions defined by function values at the vertices of a triangulation $T = (V, E, F)$ and linearly interpolated on edges and faces of T. For the moment, let us assume that no two neighboring vertices have equal function value. To succinctly describe pl-extensions of Morse theory we first need some handy definitions. Given a vertex $v \in V$ the *star* $St(v)$ is defined as the collection of simplices containing v: $St(v) = \{\sigma \in T | v \in \sigma\}$. The *upper star* $St^+(v)$ consists of all simplices in the star whose vertices all have function values higher than v: $St^+(v) = \{\sigma \in T | u \in \sigma \Rightarrow f(u) > f(v)\}$ and the *lower star* $St_-(v)$ is defined symmetrically. The *link* $Lk(v)$ is defined as the boundary of the star $Lk(v) = \partial St(v)$ and the *lower link* $LL(v)$ as the subset with function values below that of vertex v: $LL(v) = \{\sigma \in Lk(v) | u \in \sigma \Rightarrow f(u) < f(v)\}$.

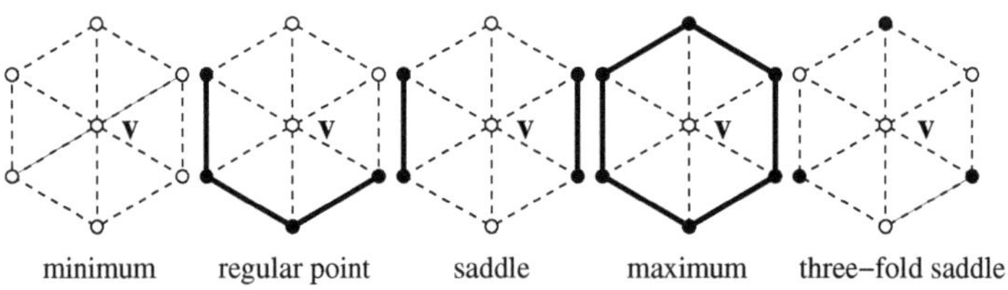

Fig. 1. Classification of a vertex v based on its lower link drawn in bold solid black

Many of the key-concepts necessary to define the Morse-Smale complex do not exist for pl-functions. Most notably, derivatives are not defined. In the following we show how all necessary concepts can be adapted to pl-functions leading to the definition of a *(quasi) Morse-Smale complex* [9]. First, we justify the assumption that no two vertices of T have the same function value by breaking ties using, for example, vertex indices. Now, the topology of the lower link can be used to classify vertices, see Fig. 1. A vertex v is a maximum if its lower link is the entire link and a minimum if its lower link is empty. In all other cases the lower link consists of $k \geq 1$ connected pieces. A vertex is regular if $k = 1$ and a k-fold saddle otherwise. Essentially, v is classified using an arbitrarily small neighborhood around v rather than derivative information at v. The next step is to ensure that there exist no multiple saddles ($k > 1$) by splitting them into simple ones as described in Sect. 3.2. Finally, integral lines are replaced by *steepest lines*. As the name suggests, steepest lines are defined as greedily following a steepest ascending or descending line starting at a given vertex. In general, steepest lines are not uniquely defined. However, given any set of non-crossing steepest lines one can define ascending and descending manifolds for pl-functions and their intersection returns a quasi MS complex [9]. A quasi MS complex is guaranteed to have the same combinatorial structure as a Morse-Smale complex but is not unique for a given pl-function. Nevertheless, the differences between two equally valid quasi MS complexes are usually minor and mostly due to inadequate sampling. In the following we ignore these differences and simply refer to the MS complex.

3 Computation

This section contains an in-depth description of how to compute MS complexes on triangulated two-manifolds with boundary. First, we introduce the data structure and list all flags needed during the algorithm. Second, we discuss the necessary algorithms in detail and provide corresponding pseudo-code to allow easy reimplementation.

3.1 Setup

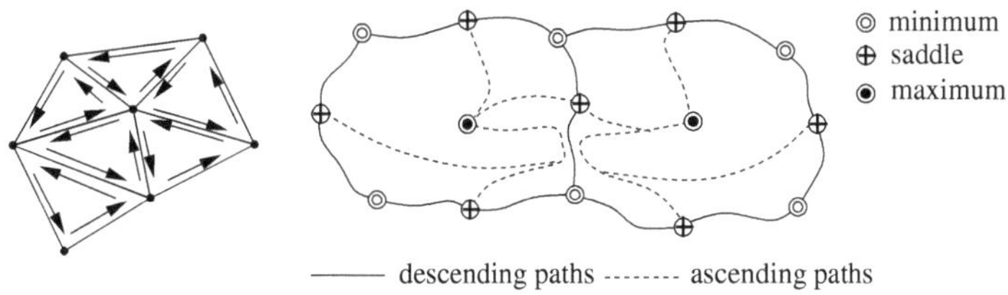

(a) (b)

Fig. 2. (**a**) The mesh is stored as a standard half-edge data structure. (**b**) If descending paths are computed first an edge can be shared by at most three paths resulting in an ascending-descending-ascending classification. Any additional path merges with one of the existing ones

We store the triangulation in a typical half-edge data structure consisting of vertices, (half-)edges, and faces, see Fig. 2(a). Each vertex stores a function value and the function is continued to edges and faces by piece-wise linear interpolation. To each element we add some flags necessary for the algorithm:

- vertex: each vertex stores a type and a classification
- half-edge: each half-edge stores a classification and direction.

A vertex type can be either minimum, maximum, saddle, or regular. Vertices and half-edges are classified as either free or part of an ascending and/or descending path. The classifications of two neighboring half-edges provide a directed classification of edges. Given a certain orientation an edge can be, for example, part of an ascending path on its left but part of a descending paths on its right. As will be discussed below we compute all descending paths before computing ascending paths and paths in the same direction merge. As a result the most complicated edge classification possible is ascending-descending-ascending as shown in Fig. 2(b). This makes it possible to store all possible edge-classifications using two bits in each half-edge. Another bit is used to indicate the direction (upwards or downwards) of a half-edge. The vertex type and classification also each take two bits. If necessary, the memory foot-print can be optimized further by encoding the edge classification in three bits per edge (not half-edge) leaving one bit for the direction. The resulting two bits per half-edge can be stored in one byte per triangle.

3.2 Algorithm

The algorithm proceeds in four steps:

1. Preparing the mesh;
2. Computing descending paths;
3. Computing ascending paths;
4. Extracting connectivity information between Morse-Smale regions.

Mesh preparation. Section 2.2 discussed how areas of zero gradient are simulated as non-degenerate by breaking ties arbitrarily via the direction flags. There exist other degeneracies that for various reasons we also resolve symbolically before computing an MS complex. In particular, we split multiple saddles into a collection of simple saddles and remove boundary saddles.

Figure 3(a) shows how a saddle s of arbitrary multiplicity can be recursively split into a collection of simple saddles. While the multiplicity of s is larger than two we compute a single component C of its lower link and create a new vertex u with $f(u) = f(s)$. We delete all edges incident to C from s and connect them to u. Furthermore, we create three new edges connecting u to s and to the two vertices a and b neighboring C in the link of s (creating two new faces in the process). Finally, we mark the (directed) edge from u to s as descending and the edges from u to a and b as ascending. As a result, u is a simple saddle and the multiplicity of s has been reduced by one. In practice, one often wants the new vertices created in the process of splitting multiple saddles to have a reasonable embedding. One choice of embedding for u that works well in practice is to pick u as the mid-point of one of the edges between s and a vertex in C, as shown in Fig. 3(a).

Other than areas of zero gradient and multiple saddles, boundary saddles are in principle not degenerate. Even smooth Morse functions on manifolds with boundary can have saddles on the boundary. However, Morse theory in

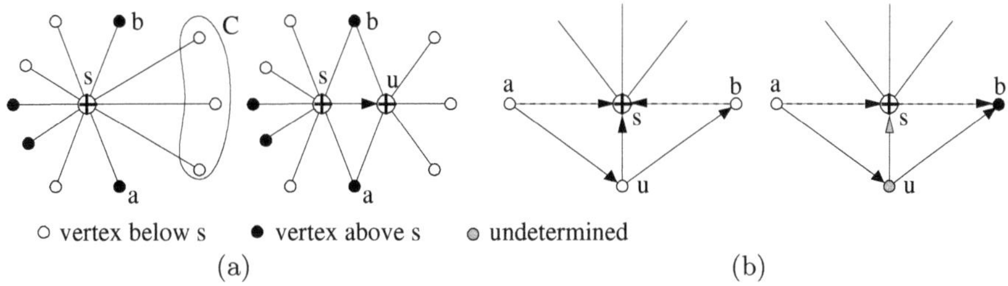

Fig. 3. Preparing the mesh: Filled/empty circles indicate vertices above/below s and for selected edges the arrow indicates ascending direction. (**a**) Splitting a multiple saddle by "removing" one component of its lower link. (**b**) Padding boundary saddles: Restricted to the boundary the saddle can be an extremum (left) or a regular vertex (right). It is padded by two triangles moving it into the interior. The grey color indicates that the choice of direction for u, s on the far right depends on the actual function values

general does not apply to manifolds with boundary. Instead, one must look to stratified Morse theory [10] for theoretically applicable results. A central assumption in stratified Morse theory is that there exist no boundary saddles. Apart from the theoretical justification for removing boundary saddles we found that they significantly complicate the computation as well as simplification of an MS complex.

We remove boundary saddles by padding them with two new triangles effectively moving them to the interior of the mesh. Figure 3(b) illustrates this process. Let s be the boundary saddle and a and b its neighboring vertices along the boundary. We create a new vertex u and two new triangles (u, s, a) and (u, b, s) and wlg. assume $f(a) < f(s)$ and $f(a) < f(b)$. We set $f(u) = 0.5 * (f(a) + f(b))$. We mark the edges (a, u) and (u, b) upward and depending on $f(u) < f(s)$ or $f(u) > f(s)$ we mark (s, u) downward or upward. As a result, u is a regular boundary vertex and s has become an interior saddle which, if necessary, can be split into simple saddles as described above. Splitting saddles as well as moving boundary saddles is a local operation that is performed during the initial read-in of the mesh and therefore does not require a dynamic mesh structure.

Descending Path Computation. For the remainder of the paper we assume that the mesh has been prepared according to the previous section and there exist no multiple or boundary saddles. In the next step we compute all descending paths. When computing integral lines in general there exist two choices. One can restrict steepest lines to follow edges in the triangulation [9] or trace steepest lines through triangles [3]. This paper focuses on the first option as it requires more complicated algorithms and potentially creates a number of challenging degenerate cases. Furthermore, tracing along existing edges is faster and requires no additional mesh refinement. Nevertheless, we indicate how all algorithms can be adapted to tracing steepest lines through triangles and what the advantages and disadvantages are in either case.

Paths are traced starting at saddles and ending at extrema. We store paths by marking edges and vertices incident to them. As discussed above, edges are marked in a directed way such that an ascending-descending classification can be distinguished from a descending-ascending one. Initially, only critical points are classified non-free: Maxima/minima are classified as part of an ascending/descending path and saddles as both.

Descending paths are computed following three rules:

R1: Two paths of the same type can merge;
R2: Paths are not allowed to split;
R3: Paths that reach the boundary stay on the boundary;

R1 is a direct consequence of f being piece-wise linear. One can easily imagine a valley floor consisting of only a single edge. All descending lines flowing into such a valley must merge on this edge. R2 is fulfilled implicitly, since each path is computed following the same (deterministic) algorithm. R3 potentially forces a path from its "correct" location since boundary edges are not necessarily the steepest edges. However, by enforcing rule R3 we effectively avoid

```
Let T = (V, E, F) be a triangulation
ComputeAllDescendingPaths(Vertices V,Edges E)
  forall s ∈ V with MorseIndex(s) == 1          //for all saddles
    forall C component of LL(s)        //for each component of the lower link
      //find the steepest edge
      u = EndPointOfSteepestDescendingEdge(C, V, E);
      ClassifyAsDescending(Edge(s, u),left&right);
      if IsDescending(u) == false        //if we have not hit an existing path
        ClassifyAsDescending(u);
        TraceSteepestDescendingLine(u);        //continue tracing
      endif
    endfor
  endfor

TraceSteepestDescendingLine(Vertex v)
  if IsBoundaryVertex(v) == true        //If v lies on the boundary
    u = GetLowerIncidentBoundaryVertex(v);        //stay on the boundary
  else
    u = EndPointOfSteepestDescendingEdge(LL(v), V, E);
  endif
  ClassifyAsDescending(Edge(s, u),left&right);
  if IsDescending(u) == false        //if we have not hit an existing path
    ClassifyAsDescending(u);
    TraceSteepestDescendingLine(u);        //continue tracing
  endif
```

Fig. 4. Pseudo-code to compute all descending paths in a triangulation

detachment points [13] and simplify constructing the final MS complex significantly. If necessary, one can check whether rule R3 has significantly altered the final complex by re-computing an unconstrained steepest line and comparing the end-points. In such a case the complex can be corrected accordingly [9].

Following these rules, computing descending paths is straightforward: In each component of the lower link of all saddles we search for a steepest descending edge and classify it descending on both sides. If its end-point a is not yet classified descending we mark it as such and recursively look for a steepest descending edge out of a. The detailed algorithm is shown in Fig. 4.

Note, that in the boundary case of TRACESTEEPESTDESCENDINGLINE there exists a unique lower boundary vertex neighboring v since otherwise v would be a saddle. Furthermore, the test whether a vertex u is already classified as descending combines three possible cases. We stop tracing a path when we have reached: another path (when we merge), a minimum (when we are done), or another saddle. In the last case the path merges with one of the descending path starting at u. However, there exist two equally valid choices. By convention, we assume that a descending path hitting another saddle verges to the right relative to its downward direction. Figure 5(a) shows a stable manifold whose boundary is computed by the algorithm described above.

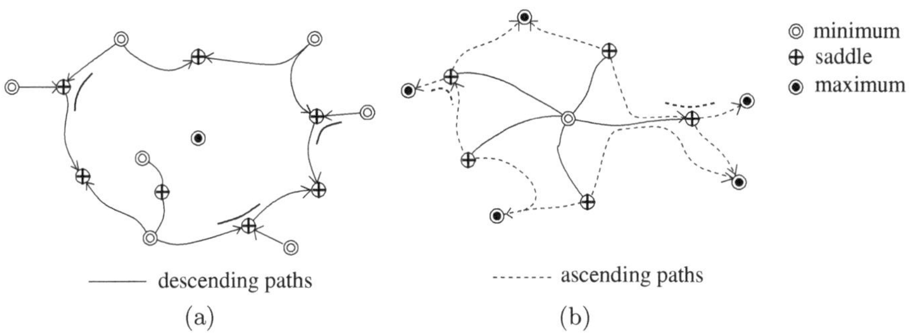

Fig. 5. (**a**) Stable manifold as computed by COMPUTEALLDESCENDINGPATHS. The arrows indicate the ascending direction and the bold line elements indicate how paths hitting saddles are routed. (**b**) Unstable manifold as computed by COMPUTEALLASCENDINGPATHS

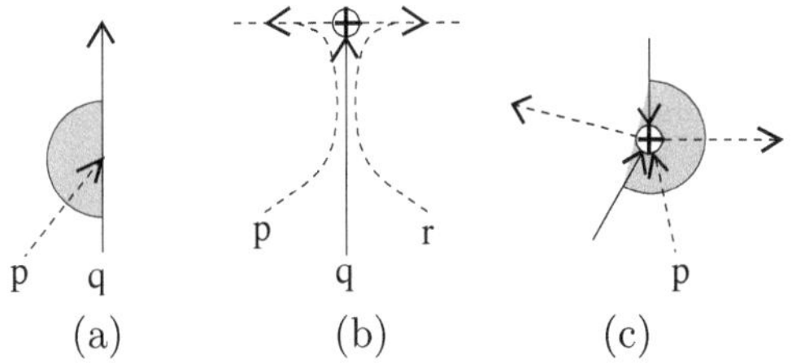

Fig. 6. Degeneracies encountered when computing ascending paths. The shaded disk slices indicate possible choices for the steepest edge.

When one is interested in tracing paths through triangles the simplest method is to change EndPointOfSteepestDescendingEdge. Rather than only examining all existing edges one computes the true steepest path. If this path runs through a triangle one splits the triangle to create a corresponding edge and returns the newly created vertex. However, refining extremely large meshes is often undesirable and in the case of out-of-core meshes can be prohibitively expensive.

Ascending path computation. In principle, ascending paths are computed symmetrically to the descending ones. The same three rules R1-R3 hold but another one must be added:

R4: Two paths are not allowed to cross.

Whenever one looks to extend a steepest ascending path the choice of "steepest" edge must be modified to observe rule R4. This becomes relevant when descending and ascending paths share a vertex. Figure 6 shows some examples in which special care must be taken to not violate R4. In Fig. 6(a) an ascending path p has joined a descending path q from the left (relative to p's direction). Since p must not cross q only those ascending edges to the left of q or part of q can be used to extend p. In particular, there can exist another ascending path r joining q from the right as shown in Fig. 6(b). Even though p and r share edges they are separated by q and cannot merge. If they would be allowed to merge they could never split (rule R2) and therefore one of them would cross q at some point. The same principles apply if an ascending paths p hits a saddle s, see Fig. 6(c). There can exist only one ascending path starting at s that p can join without crossing descending paths. Given these conventions, the ascending paths can be computed as shown in Fig. 7.

There are two significant differences between computing the ascending compared to the descending paths. First, an ascending path does not necessarily stop once it hits another vertex classified as ascending or even a saddle. Only reaching a maximum or sharing an edge already marked (on the correct side) is a sufficient condition to stop tracing. Second, the search for steepest edges is constrained by rule $R4$ and therefore only considers parts of the upper star of a vertex, see Fig. 6. As before, the function SteepestAscendingEdge can be adapted to compute the true steepest ascent by splitting triangles. Furthermore, it can be modified to refine the mesh to keep ascending and descending paths completely separate. The disadvantage is again the potential cost of refining the mesh, the advantage is that extracting the connectivity between MS regions, see below, becomes trivial.

Extracting connectivity. Once all paths are constructed one also knows all arcs (endpoints and geometric embedding) of the MS complex. Then we identify the starting edge of each path with the corresponding arc. (These are the edges used for the first step in ComputeAll(A-)DescendingPaths.) Trivially, the order of arcs around saddles is given by the order of these edges. An additional breadth-first traversal of only descending paths starting from the

```
Let T = (V, E, F) be a triangulation
ComputeAllAscendingPaths(Vertices V, Edges E)
  forall s ∈ V with MorseIndex(s) == 1          //for all saddles
    forall C component of St+(s)       //for each component of the upper star
      e = SteepestAscendingEdge(C, V, E);        //find the steepest edge
      //now we make an arbitrary choice to start all
      //ascending paths on the right side of an edge
      ClassifyAsAscending(e,right);
      if MorseIndex(v) ≠ 2  AND  IsAscending(u,right) = false
        TraceSteepestAscendingLine(e,right);        //start tracing
      endif
    endfor
  endfor

TraceSteepestAscendingLine(Edge e, Orientation side)
  v = Endpoint(e);
  if IsBoundaryVertex(v) = true        //If v lies on the boundary
    u = AscendingBoundaryEdge(v);
  else if IsDescending(v) = true  AND  side = right
    u = SteepestAscendingEdge(RightHalf(St+(v)),V, E);
  else if IsDescending(v) = true  AND  side = left
    u = SteepestAscendingEdge(LeftHalf(St+(v)),V, E);
  else
    u = SteepestAscendingEdge(St+(v), V, E);
  endif
  v = Endpoint(u);
  if MorseIndex(v) ≠ 2  AND  IsAscending(u,side) = false
    ClassifyAsAscending(u,side);
    if IsDescending(u,side)
      TraceSteepestAscendingLine(u,side);
    else if IsDescending(v) = true  AND   u ∈ LeftHalf(St−(v))
      TraceSteepestAscendingLine(u,left);
    else
      TraceSteepestAscendingLine(u,right);
    endif
  endif
```

Fig. 7. Pseudo-code to compute all ascending paths in a triangulation

minima determines the order of arcs around minima. Note, that care must be taken for descending paths flowing though saddles. Here, observing the earlier convention that paths always turn right at saddles is crucial. Knowing the order of arcs around minima and saddles is enough to deduce the order of arcs around maxima which completes the connectivity information. If necessary one can perform a flood-fill to determine which faces belong to which Morse-Smale region. When ascending and descending paths are kept disjunct the connectivity can also be computed using a simple flood-fill algorithm. In this case each Morse-Smale region corresponds to a single cluster of faces which form a topological disc. The clusters must touch all four arcs of the region and the connectivity is induced directly by the connectivity of the mesh.

4 Simplification

In most applications computing the topology of a data set is only the first step. To use the topology for analysis one must often simplify it if only to remove noise. This section describes how to store an MS complex as a progressive mesh [15]. A progressive mesh is a data structure that for a given error threshold allows fast access to a simplified MS complex in which all features below the error threshold have been removed.

The generic operation to simplify an MS complex is called a *cancellation.* An example is shown in Fig. 8(a) and (b). The maximum-saddle pair u, v

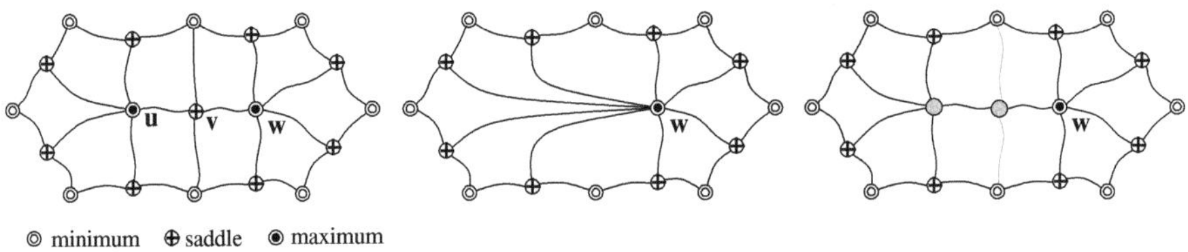

(a) (b) (c)

Fig. 8. Situation before (**a**) and after (**b**) canceling the critical point pair u, v. (**c**) Implementing a cancellation by deactivating nodes and arcs

is removed by merging four regions into two and extending all arcs ending at u. The reverse operation which introduces u, v and splits the regions is called an *anti-cancellation.* To rank cancellations each one is assigned an error indicating the importance of the corresponding node pair. Some examples are the difference in function value between the critical points involved (also called persistence), geometric distance, or absolute function value.

Valid cancellations. Not all connected critical point pairs can be canceled. For example, if a saddle is connected to the same extremum twice, see Fig. 9(a), it cannot be canceled with this extremum. Such a cancellation would require a change in genus of the underlying manifold which is usually not desirable. For MS complexes without boundary such double-connected saddles are the only configuration which cannot be canceled. For manifolds with boundary, however, there exist additional rules. Figure 9(b) shows why a boundary extremum can only be canceled if its "opposite" extremum across the saddle also lies on the boundary. Canceling a boundary extremum to the inside would create a boundary saddle which is disallowed. Let us assume a saddle is connected to two boundary maxima and should be canceled with either of them. The boundary minimum between the two maxima is implicitly

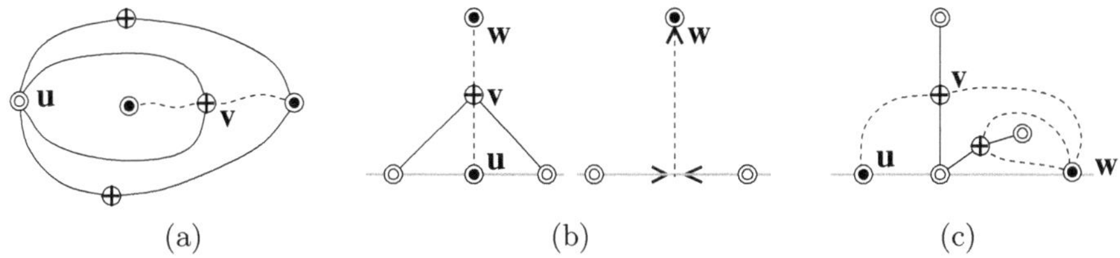

Fig. 9. Invalid cancellations: (**a**) Canceling u, v would change the genus of the underlying manifold. (**b**) Canceling u, v creates a boundary saddle at u. (**c**) Canceling u, v would remove all critical points between $u - v - w$ and the boundary

removed as well since the restriction of f to the boundary is a one-dimensional function of which minima and maxima can only be canceled in pairs. However, if the minimum is connected to additional saddles, as shown in Fig. 9(c), these structures must be removed as well since without the minimum they cannot form part of a valid MS complex. To avoid such implicit multi-cancellations we disallow the cancellation of boundary extrema unless they remove exactly three critical points. This condition is equivalent to requiring the minimum between the two maxima in our example to have valence one. Subsequently, we call a cancellation *valid* if it does not violate one of the three rules described above.

Construction. Given an error metric it is easy to progressively simplify an MS complex. Each arc of the MS complex corresponds to a possible (but not necessarily valid) cancellation. All valid cancellations are added into a priority queue which returns the arc/cancellation with the smallest error. As long as the queue is not empty we greedily cancel the critical point pair with smallest error. Note, that performing one cancellation can invalidate other cancellations. For example, a saddle can usually be canceled either with a minimum or with a maximum and clearly either choice will invalidate the other. For most metrics the error of a cancellation can be changed by neighboring cancellations as the connectivity of the complex changes. Typically, however, the error is only increased by earlier cancellations and therefore it can be updated in a lazy fashion. The complete simplification algorithm is shown in Fig. 10.

4.1 Progressive MS Complexes

The algorithm of Fig. 10 simplifies an MS complex until no valid cancellations remain. By introducing a break-off point it can easily be adapted to simplify a complex up to a certain error threshold. In practice, however, one usually wants to interactively change this threshold and simplifying the complex in a bottom-up fashion each time is slow and cumbersome. Instead, one wants to maximally simplify the complex once and keep a record of all intermediate configurations.

We store an MS complex as a graph consisting of nodes and arcs. Arcs are assumed to be ordered around nodes and regions are not represented explicitly.

```
Let MS = (N, A, R) be an MS complex
Let Q be a priority queue of arcs ordered by increasing error
SIMPLIFY(MS complex MS)
  forall a ∈ A     //for all arcs in the complex
    if ISVALIDCANCELLATION(a)
      PUSH(Q, a, COMPUTEERROR(a));
    endif
  endfor
  while ISEMPTY(Q) == false
    top = POP(Q);
    if ISVALIDCANCELLATION(top)     // if this cancellation is still valid
      err = COMPUTEERROR(top);
      if (err ≤ NEXTPRIORITY(Q);
        CANCELCRITICALPOINTS(a);
      else
        PUSH(Q, a, err);
      endif
    endif
  endwhile
```

Fig. 10. Pseudo-code to maximally simplify an MS complex

Using this simple data structure we implement a cancellation as shown in Fig. 8(c). Rather than deleting the two arcs and two critical points, we only "deactivate" them. deactivating a node creates "super-arcs" by concatenating the remaining active arcs and deactivated arcs are treated as being removed. A progressive MS complex is stored by assigning each node and each arc an error. Prior to simplification, the errors are initialized to a value larger than the largest possible error. During a cancellation elements are deactivated by setting their error to the error of the current cancellation. For any given error threshold the corresponding simplified MS complex is given by the collection of active nodes and super-arcs, where an element is considered active if its error is larger than the current threshold.

To explicitly construct a simplified MS complex for a particular error budget one performs restricted breadth-first traversals of the graph starting at each active extremum. The restrictions are that only active arcs and deactivated nodes can can be traversed. Each traversal covers a tree of active arcs rooted at the starting extremum with either active saddles or deactivated extrema as leaves. (Note, that an active saddle can be a leaf of two different branches of this tree. However, in this case it makes sense to consider the saddle as two nodes and maintain that the structure has no cycle.) Each path from a leaf to the root represents a super-arc. The simplified MS complex consists of all active nodes and those super-arcs that start at saddles, see Fig. 11. In practice, the tree traversals are helpful in a variety of situations. For example, a complex can be rendered very efficiently by associating each arc with a line strip and creating appropriate display lists during the tree traversals. Furthermore, when one considers a cancellation as the merging of

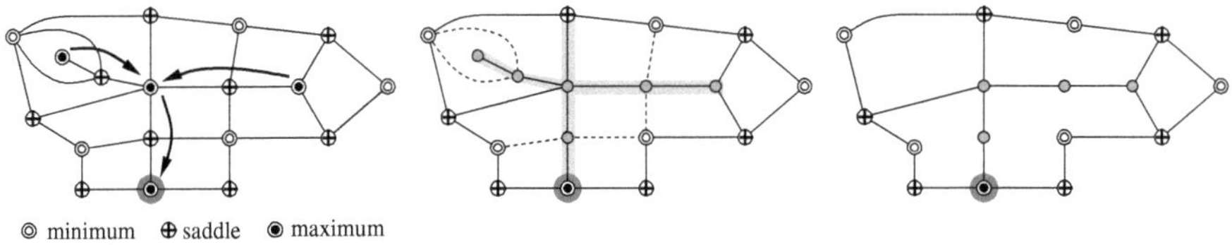

(a) (b) (c)

Fig. 11. Accessing a simplified MS complex using tree traversals. (**a**) Original MS complex with three cancellations indicated by the arrows. (**b**) The complex of (a) after the cancellations. The tree traversed from the indicated root is shaded in grey. (**c**) The final MS complex. Note, that the left most branch of the tree ended at an extremum and has been removed

three critical points into one, the deactivated nodes in each tree represent all critical points merged with the root. The main advantage of this data structure is that it is small and very easy to implement. The only disadvantage is that creating a simplification takes time linear in the size of the full complex, rather than linear in the size of the simplification. However, in all applications we have encountered so far the MS complex is several orders of magnitude smaller than the surfaces mesh. For example, the full resolution surfaces of the Mixing Fluids data set (see Sect. 5) contain up to 22 million vertices yet never more than about 20.000 maxima/bubbles. Therefore, the simplification remains fully interactive for all models we have been able to process using this approach.

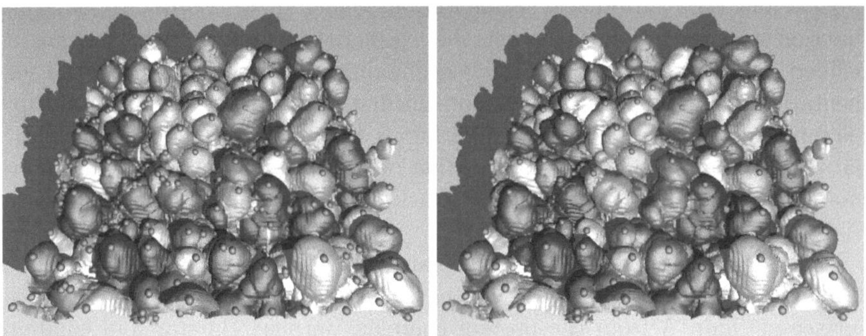

(a) (b)

Fig. 12. Segmentation of parts of the Mixing Fluids data sets into bubbles. Maxima are shown in red and each bubble is randomly assigned one of nine colors. (**a**) Initial segmentation; (**b**) Segmentation at 0.2% persistence (relative to the maximal range in function value). (Colorplate on p. 218.)

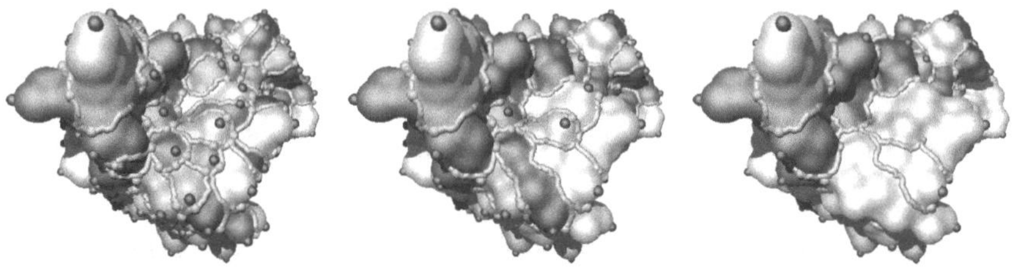

(a) (b) (c)

Fig. 13. Segmentation of the atomic density function on a molecule. Minima are shown in blue and ascending paths in gold. Segmentation into 198 (**a**); 100 (**b**); and 50 (**c**) protrusions (colorplate on p. 218).

5 Applications

The algorithms described above provide an efficient and stable method to compute MS complexes of any function on a triangulated manifold. We have found this a very versatile tool helpful in a variety of independent areas. Here, we show three very different applications for which we currently use MS complexes: The first is a physics simulation of the turbulent mixing between two fluids. Figure 12 shows an iso-surface between two mixing fluids extracted from one time-step of a simulation performed at the Lawrence Livermore National Laboratory. The data has been generated by the Miranda code a higher order hydrodynamics code for computing fluid instabilities and turbulent mixing [6]. In particular, scientists are interested in "bubbles" formed during the mixing process and their automatic segmentation. Using the z-coordinate as Morse function and the iso-surface (not the xy-plane) as domain bubbles can be defined as the descending manifolds of maxima. Nevertheless, the initial segmentation shown in Fig. 12(a) is not optimal as some bubbles have multiple maxima and there exist many superfluous maxima caused by noise in the data set. Using a uniform simplification of the MS complex one can remove most of these artifacts and create a much cleaner segmentation, as shown in Fig. 12(b).

The second application is molecular biology where one is interested in segmenting a molecular surface into cavities and protrusions. We take a skin surface of chain A from the protein complex Barnase/Barstar and compute the atomic density function over this surface. The ascending manifolds of minima of this function segment the surface into protrusions, see Fig. 13. As with bubbles, simplifying the MS complex captures protrusions at coarser and coarser level.

Finally, we use the MS complex to help remesh surfaces. As described in [7], MS complexes of eigenfunctions of the Laplace matrix of a surface are well suited to form an all quadrilateral basemesh, see Fig. 14.

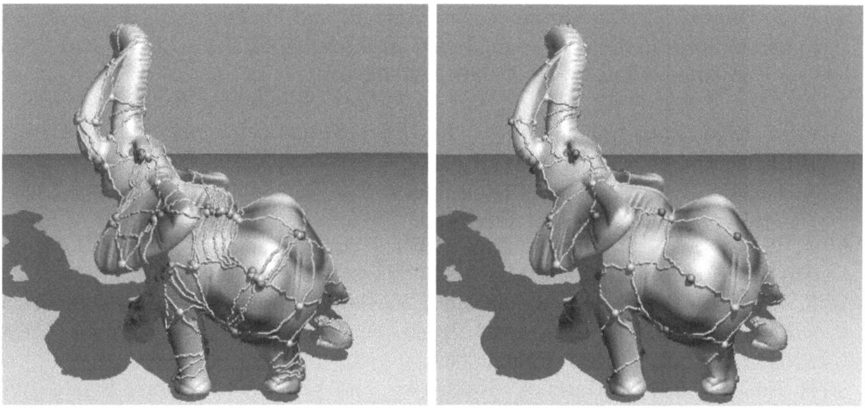

(a) (b)

Fig. 14. Creating a basemesh from Laplacian eigenfunctions. (**a**) Initial MS complex of the 60th eigenfunction of the elephant showing typical noise due to discretization and the iterative eigensolver. (**b**) The MS complex of (a) simplified to a persistence of 0.5%. All noise has been removed and the MS complex now forms a high quality all-quadrilateral basemesh (colorplate on p. 218).

6 Conclusions

We have shown a new algorithm to compute two-dimensional MS complexes which allows a straightforward implementation of the original methods discussed in [9]. Except for few cases during an initial setup phase no refinement of the mesh is necessary. Nevertheless, if desired the algorithm can be trivially adapted to the one discussed in [3] which computes true lines of steepest a-/descent. Additionally, we have provided a novel easy-to-implement data structure for MS complexes which can also be used to encode the complex progressively. Finally, we have introduced restricted tree traversals to allow fast access to a simplified complex and related information as well as to provide efficient rendering of simplified complexes.

Acknowledgments

This work was performed under the auspices of the U. S. Department of Energy by University of California Lawrence Livermore National Laboratory under contract No. W-7405-Eng-48.

References

1. C. L. Bajaj, V. Pascucci, and D. R. Schikore. Visualization of scalar topology for structural enhancement. In D. Ebert, H. Hagen, and H. Rushmeier, editors, *Proc. IEEE Visualization '98*, pages 51–58, Los Alamitos California, 1998. IEEE, IEEE Computer Society Press.
2. C. L. Bajaj and D. R. Schikore. Topology preserving data simplification with error bounds. *Computers and Graphics*, 22(1):3–12, 1998.

3. P.-T. Bremer, H. Edelsbrunner, B. Hamann, and V. Pascucci. A topological hierarchy for functions on triangulated surfaces. *IEEE Trans. on Visualization and Computer Graphics*, 10(4):385–396, 2004.
4. H. Carr, J. Snoeyink, and M. van de Panne. Simplifying flexible isosurfaces using local geometric measures. In *IEEE Visualization '04*, pages 497–504. IEEE Computer Society, 2004.
5. A. Cayley. On contour and slope lines. *The London, Edinburgh and Dublin Philosophical Magazine and Journal of Science*, XVIII:264–268, 1859.
6. A.W. Cook, W. Cabot, and P.L. Miller. The mixing transition in Rayleigh-Taylor instability. *J. Fluid Mech.*, 511:333–362, 2004.
7. S. Dong, P.-T. Bremer, M. Garland, V. Pascucci, and J. C. Hart. Spectral surface quadrangulation. *ACM Trans. on Graphics (TOG) / Proc.of ACM SIGGRAPH*, 25(3):1057–1066, 2006.
8. H. Edelsbrunner, J. Harer, V. Natarajan, and V. Pascucci. Morse-Smale complexes for piecewise linear 3-manifolds. In *Proc. 19th Sympos. Comput. Geom.*, pages 361–370, 2003.
9. H. Edelsbrunner, J. Harer, and A. Zomorodian. Hierarchical Morse-Smale complexes for piecewise linear 2-manifolds. *Discrete Comput. Geom.*, 30:87–107, 2003.
10. M. Goresky and R. MacPherson. *Stratified Morse Theory.* Springer-Verlag, Heidelberg, Germany, 1988.
11. A. Gyulassy, V. Natarajan, V. Pascucci, P.-T. Bremer, and B. Hamann. Topology-based simplification for feature extraction from 3D scalar fields. In *Proceedings of the IEEE Visualization 2005 (VIS'05)*, pages 535–542. IEEE Computer Society, 2005.
12. R. M. Haralick and L. G. Shapiro. Image segmentation techniques. *CVGIP*, 29(1):100–132, January 1985.
13. J. L. Helman and L. Hesselink. Visualizing vector field topology in fluid flows. *IEEE Computer Graphics and Applications*, 11(3):36–46, May/Jun. 1991.
14. M. Hilaga, Y. Shinagawa, T. Kohmura, and T. L. Kunii. Topology matching for fully automatic similarity estimation of 3D shapes. In E. Fiume, editor, *Proceedings of ACM SIGGRAPH 2001*, pages 203–212, New York, NY, USA, 2001. ACM.
15. H. Hoppe. Progressive meshes. *Computer Graphics (Proc. SIGGRAPH '96)*, 30(4):99–108, Aug. 1996.
16. Y. Matsumoto. *An Introduction to Morse Theory.* American Mathematical Society, 2002.
17. J. C. Maxwell. On hills and dales. *The London, Edinburgh and Dublin Philosophical Magazine and Journal of Science*, XL:421–427, 1870.
18. J. Milnor. *Morse Theory.* Princeton University Press, New Jersey, 1963.
19. L. R. Nackman. Two-dimensional critical point configuration graphs. *IEEE Transactions on Pattern Analysis and Machine Intelligence*, 6(4):442–450, July 1984.
20. J. Pfaltz. Surface networks. *Geographical Analysis*, 8:77–93, 1976.
21. J. Pfaltz. A graph grammar that describes the set of two-dimensional surface networks. *Graph-Grammars and Their Application to Computer Science and Biology (Lecture Notes in Computer Science*, no. 73, 1979.
22. Y. Shinagawa, T. Kunii, and Y. L. Kergosien. Surface coding based on Morse theory. *IEEE Computer Graphics and Applications*, 11:66–78, 1991.
23. S. Takahashi, T. Ikeda, Y. Shinigawa, T. L. Kunii, and M. Ueda. Algorithms for extracting correct critical points and constructing topological graphs from discrete geographical elevation data. In *Proc. Eurographics '95*, pages C–181–C–192, Sep. 1995.

24. S. Takahashi, Y. Shinagawa, and T. L. Kunii. A feature-based approach for smooth surfaces. In *Proc. of the Fourth Symp on Solid Modeling and Applications, SMA '97*, pages 97–110. ACM, May 1997.
25. M. J. van Kreveld, R. van Oostrum, C. L. Bajaj, V. Pascucci, and D. Schikore. Contour trees and small seed sets for isosurface traversal. In *Symposium on Computational Geometry*, pages 212–220, 1997.
26. W. Warntz. The topology of a socio-economic terrain and spatial flows. *Regional Scientific Associations*, 17:47–61, 1966.

On the Role of Topology in Focus+Context Visualization

Ivan Viola and Eduard Gröller

[1] Department of Informatics, University of Bergen, Norway
ivan@ii.uib.no
[2] Institute of Computer Graphics and Algorithms,
Vienna University of Technology, Austria
meister@cg.tuwien.ac.at

Summary. In this paper three types of visualization scenarios are discussed, where topology improves the readability of particular visualization results. The first type combines topology information represented by simple graphical primitives with other forms of visual representations. The second type uses the topology information to define the relevance of objects within the data. The relevance is reflected in the visualization by applying the cut-away concept. The third type of visualizations is based on the change of topology of the underlying data to increase visibility of the most interesting information. Every type handles topology in a different way. This illustrates various roles of topology in scientific visualization.

1 Introduction

Three-dimensional visualization is becoming an essential tool for the analysis of various scientific problems. A major problem are the very large data sizes that are hard to handle. For example the rapid development of high-precision medical imaging modalities causes the amount of data to steadily increase. Due to the increasing power of modern CPUs, mathematical simulations of scientific phenomena deliver huge amounts of result data. Processing and visualization of time-varying data is becoming practicable in many applications. The added temporal dimension, furthermore increases the data sizes considerably.

Large data sizes entail two fundamental problems: The first is data manipulation with respect to data enhancement and processing in general. In this paper we focus on the second problem, which is the appropriate visual representation of the underlying data. The amount of relevant information is often relatively small as compared to the overall amount of acquired data. Therefore these small, interesting features have to be emphasized visually.

We generalize the term *feature* as a part of the data that is of special interest to the user. To highlight the role of topology in visualization, we

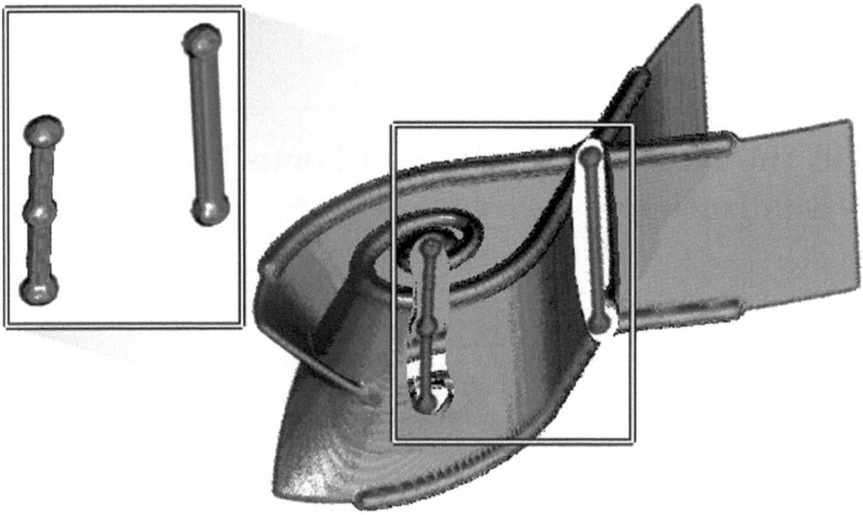

Fig. 1. Periodic Blue Sky Bifurcation: The topology can be clearly depicted by simple and sparse geometry. Critical lines occluded by separation surfaces are unveiled by using importance-driven visualization, where importance is defined through topological characteristics. Additional emphasis is achieved by displacement in the spirit of exploded views. The polygonal model is courtesy of Tino Weinkauf and Holger Theisel (colorplate on p. 219).

concentrate in our examples on features that have some topological characteristics. Examples are singularities in flow visualization, or cancer lung nodules in medical visualization.

In many cases visualization of the features alone does not clearly illustrate the inspected problem and thus does not satisfy the goal of an expressive visualization. The data surrounding the features includes more information and some of this information is necessary to communicate the visualized problem. Spatial position and vicinity to other structures can be very important for example. A visualization technique that combines the most interesting data (i.e., a feature) with surrounding information of smaller information value is in general denoted as focus+context visualization [2].

Combining the most interesting features with the contextual information has to be done carefully to avoid visual overload. In this paper we discuss several visualization scenarios, where topology helps to effectively convey the information in focus+context visualizations:

- **Visualization of topology**: An appropriate visual representation is the key to an effective focus+context visualization. The general idea is to provide visualizations as simple as possible to clearly describe the data. Simple spheres, lines, arrows or other basic geometric elements are used to describe abstract characteristics of the data such as topology. This abstract visual representation is very sparse, but also very clear and easily understandable. In focus+context visualization such information is combined with denser visual representations that describe the data in more detail. An abstract visual representation depicting the topology is often used in flow visualization or visualization of complex dynamical systems. Specific examples will be discussed in more detail in Section 2.

- **Topology-driven visualization**: Features within the data can be characterized by various properties, e.g., by their topological characteristics. Such characteristics can be described by very simple geometry, which can be used as an underlying structure that drives the visualization. This structure is used for example in visualization to guarantee the visibility of important features. Specific examples will be given in Section 3.
- **Topology changes of the underlying data**: The original arrangement of structures in the data is often not optimal to achieve a particular goal through visualization. It is sometimes easier to change the topology of the data in order to clearly see particular features. Examples of changing the topology (e.g., from cylindrical to planar) will be presented in Section 4.

A visualization example illustrating the role of topology in visualization is shown in Figure 1. Here all three scenarios are combined together. The Periodic Blue Sky Bifurcation [7] is represented sparsely by simple geometric elements. The goal is to visualize all topological structures, i.e., critical points, critical lines, and separatrices. The visualization is enhanced by cut-away views and a placement in spirit of exploded views to focus the attention to critical lines and points. After incorporating cut-away views the visual presence of these small features is more balanced with visually more prominent separatrices. Thus the overall visual information is increased.

In the next sections we will explain the relevance of topology in the visualization for specific applications.

2 Visualization of Topology

The relevance of features in an image is given through their visual representations, which may vary from an abstract to a more direct visual representation. A direct representation does not use any abstraction and is easy to interpret. However for a large dataset or in higher dimensions it easily produces dense, cluttered images that are visually overloaded. The other end of the spectrum are highly abstract (derived) data representations. They typically produce rather sparse images but can require quite some effort to understand. Showing MRI data of the brain as greyscale images is a rather direct representation. Showing the same data through diffusion tensor imaging is a more abstract representation. Another example of a direct representation are satellite photographs. A map with city icons scaled by population including labels and connected by transport infrastructure is an abstract representation. There are many ways to design appropriate visual representations. We will focus

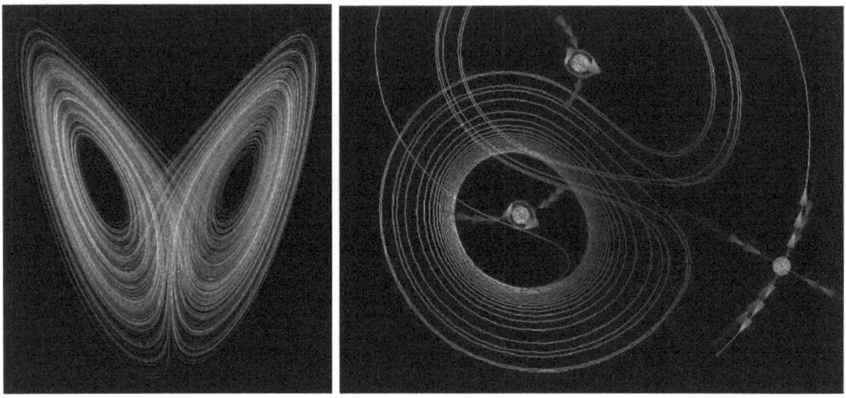

(a) (b)

Fig. 2. Lorenz attractor: (a) direct representation through a streamline and (b) descriptive visualization of the system with emphasis of critical points [4] (colorplate on p. 219).

on those where the visualization of topology significantly increases the visual information content.

The visual encoding of topology information may consist of just a few points, lines, sometimes surfaces. In cases where a direct visual representation would lead to visual clutter, the visualization of topology is a very effective way to show the most important information of the data. There is no visual overload and image space is saved to display further information. In the following we discuss two application domains where the visualization of topology efficiently conveys the characteristics of the underlying data.

2.1 Visualization of Complex Dynamical Systems

Topological visual representations are well known from complex dynamical systems. The direct representation of a complex dynamical system can lead to a visual overload. Visualization of topology on the other hand allows to focus the attention to the most relevant features. Here the visual emphasis of the most relevant structures is analogous to the difference between an aerial photograph and a geographic map.

In Figure 2 the Lorenz attractor is represented by (a) a simple streamline and (b) topological information of the underlying dynamical system. The first representation is depicting the overall shape of the system whereas the visualization of topology emphasizes critical points of the system. This information *explains* the behavior of the dynamical system more effectively. For example the alternating dominance of two critical points is caused by the third critical point. This information can by hardly extracted from the streamline plot only.

2.2 Visualization of Flow

Flow visualization is another application domain where topological visual representations are effectively used. In some cases features are not clearly defined, i.e., there are no explicit boundaries between features. Interesting features might be vortex cores, fixed points (such as nodes, saddles, vortices), separatrices etc. For a clear visualization of a particular phenomena, the topological information of interesting features is crucial. Figure 3 shows an example of a focus+context visualization of a three-dimensional flow. Here various levels of abstraction in the visual representation are present. The topological information is given by a cycle around the z-axis. The flow is directly represented by spot noise [9] on the Poincaré cross section [4]. A streamsurface in orange and a streamline in green provide further abstract context information. The sparse topological information indicates the overall characteristics of the system. The dense and direct depiction of the flow is limited to just a small region of the data to avoid visual clutter. Further examples of topological visual representations of flow can be found in related works [6, 8, 13, 14].

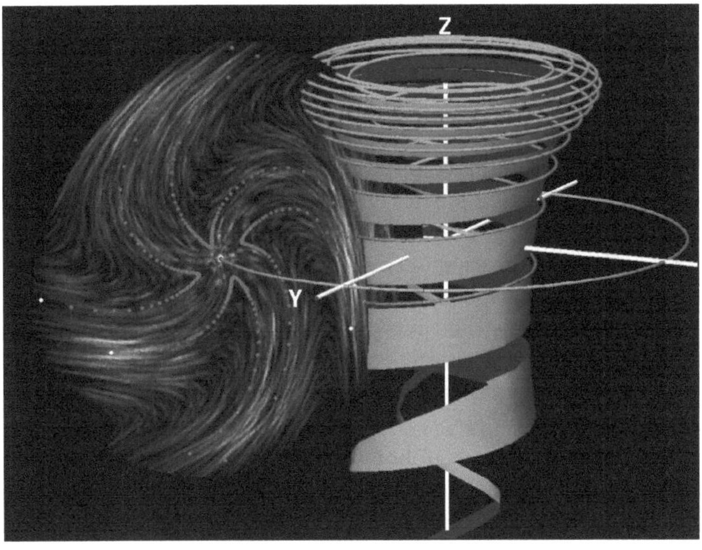

Fig. 3. Focus+context visualization of flow topology using Poincaré sections [4].

3 Topology-Driven Visualization

The previous section addressed visualizations where topological characteristics increase the visual content despite the fact that this information is represented very sparsely. In this section we will focus on visualization techniques where the topological information *drives* the visualization algorithm to emphasize interesting objects.

The concept of topology-driven visualization is analogous to importance-driven feature enhancement [12]. The traditional volume visualization pipeline assigns to features optical properties like color and opacity. With importance-driven feature enhancement another dimension is used which describes the importance of interesting objects. In topology-driven visualization the importance is defined through topological characteristics that have the highest priority to be clearly visible. Prior to image synthesis, the visibility of interesting objects is estimated. If less important objects are occluding objects that are more interesting, the less important ones are rendered more sparsely, e.g., more transparently. If the same object does not cause any unwanted occlusions in other regions of the image, it is rendered more densely, e.g., opaque, in order to see its shape more clearly. This allows to see all interesting structures irrespective if they would be occluded or not, and the less important parts are still visible as much as possible. This concept has been inspired by expressive illustration techniques known as cut-aways or ghosted views. We will show their effective usage in the scope of medical and meteorological applications.

3.1 Visualization of Early-Stage Lung Cancer

The goal of visualization of early stage lung cancer is to improve the diagnosis and to increase the probability of patient survival. The automatic diagnostic process is divided into two parts. The first part is the detection stage where the data is classified according to the probability of a tumor presence. The second part is the visualization of these suspicious regions. The medical expert is guided by the application to the regions where with high probability lung nodules are present. Thus the time-consuming process of slice-by-slice inspection is replaced by an automatic process. The medical expert finally has to validate which regions are considered as nodules and which regions are false positives.

Nodules are small features in human lungs with a spherical shape. Healthy lungs feature only tubular structures of two categories: airways and blood vessels. Healthy lungs can be considered of genus zero. Each lung nodule is seen as a *topological hole* in the healthy lungs. The presence of a nodule changes the topology of the lungs and the number of nodules defines the genus.

The detection part is based on local shape properties of a sample position. The detection algorithm is in spirit of the method proposed by Sato et al. [5]. Local shape properties are derived from the Hessian matrix which consists of the second-order partial derivatives [3]. The eigenvalues of the Hessian matrix are used to determine the probability of a spherical shape (nodule). The nodule shape is identified due to its different shape characteristics as compared to other tissues within the human lungs.

The visualization part is based on importance-driven feature enhancement. Here the feature is a part of the volume that has a high probability of being a nodule, i.e., a topological hole in the healthy tissue. Volume regions that occlude these suspicious areas are represented sparsely. Visualizations show the spatial position of the most interesting regions irrespective from the viewpoint settings. This increases the expressivity as compared to traditional visualization methods (e.g., direct volume rendering or slicing). Result visualizations are shown in Figure 4.

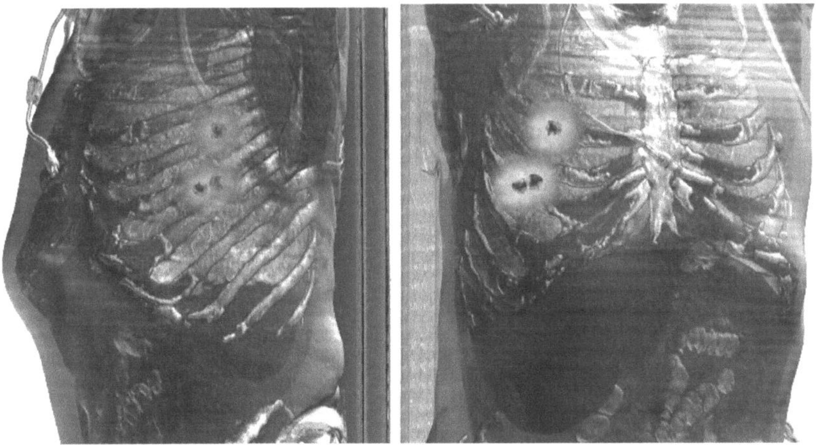

Fig. 4. Topology-driven visualizations of suspicious regions (possibly early-stage lung cancer nodules) within a thorax CT data. The suspicious regions are highlighted in red enclosed by a green halo [11] (colorplate on p. 220).

3.2 Visualization of Time-Varying Meteorological Data

The second example where topology information controls the visualization pipeline is meteorological visualization. It is illustrated by visualizing the time-varying data of a hurricane (Isabel, 2003). The data consists of multiple variables including cloud density, precipitation, ice level, temperature, pressure, and velocity vector for each time-step. In this case the interesting feature is the vortex core, i.e., the eye of the hurricane which is a topological structure. For the hurricane analysis it might be useful to have a clear view on this feature and to select which variable to visualize. Therefore the cutaway technique can be effectively applied in this case as well. The example also shows multiple variables using topology-driven visualization. The focus object is defined as the group of voxels inside a cylinder around the hurricane eye. Inside the cylinder the total precipitation mixing ratio is visualized. Due to the cut-away view it is possible to have a clear view of this property close to the eye of the hurricane. Outside the cylinder is the context area where the total cloud moisture is visualized. In this time-dependent dataset the important feature changes its position and shape over time. Importance-driven visualization guarantees to emphasize the important feature irrespective of viewpoint and feature position and shape. A single time-step is shown in Figure 5.

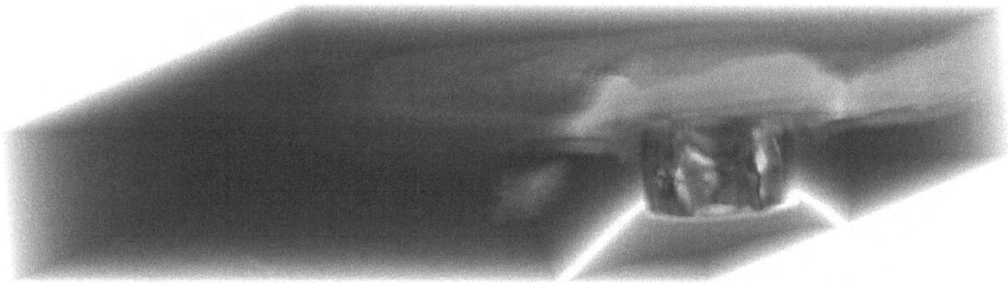

Fig. 5. Cut-away visualization of a multidimensional volumetric data of hurricane Isabel (colorplate on p. 219).

4 Topology Changes of the Underlying Data

The third type of visualizations enhances the visibility of interesting objects by (virtually) changing the topology of the underlying data. Examples include exploded views, a concept known from illustrations and assembly manuals, and other specific visualizations from computer-aided diagnosis.

4.1 Exploded Views

Exploded views modify the spatial arrangement of features to uncover the most prominent ones and thus change the topology of the underlying data. A multi-object visualization technique related to exploded views has been presented by Grimm et al. [1]. They present a data structure denoted as *V-Objects* for individual handling and manipulation of features within a volumetric dataset. An example of a topological change is shown in Figure 6. Here the context information is the skull which occludes the brain in focus. The topology of the skull is changed through the exploded views concept to unveil the focus.

4.2 Colon Unfolding

In this section we discuss virtual colon unfolding from medical visualization. The goal is to clearly present all relevant information in a single image. The

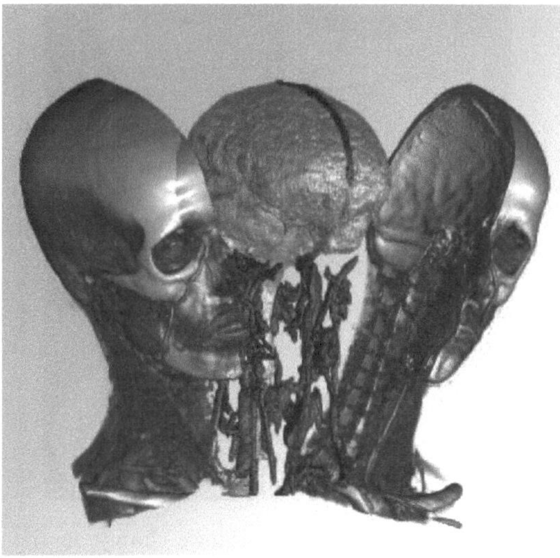

Fig. 6. Multi-volume visualization enabling exploded views through topology change of the underlying data [1] (colorplate on p. 220).

approach performs unfolding of a cylindrical structure to a plane. This incurs a topological change of the original tubular structure. The unfolding is applied to computed tomography data of the colon. The traditional way of colon inspection for polyps (early-stage colon cancer) is using a real-world endoscope. Such an inspection process is uncomfortable for the patient. Virtual colonoscopy has been applied recently. The endoscope is replaced by a virtual endoscope traversing the tomographic data instead of the patient's body. This technique is much more comfortable, but the diagnosis is still time-consuming. Unfolding the tubular structure provides an instant overview of the entire organ [10]. In this case the topological change is important for a rapid diagnosis. An image of part of an unfolded colon with identified polyps is shown in Figure 7.

5 Conclusions

In this paper we have discussed three scenarios where topology and visualization play together: visualization of topology, topology-driven visualization, and topology changes of the underlying data.

We have dealt with topology-based techniques for visualization and feature definition. The sparse visual representation showing topological characteristics is very useful for flow visualization and dynamical-systems visualization. Topological analysis can be used to define the importance of interesting objects and express their relevance in cut-away visualizations. Topological changes of

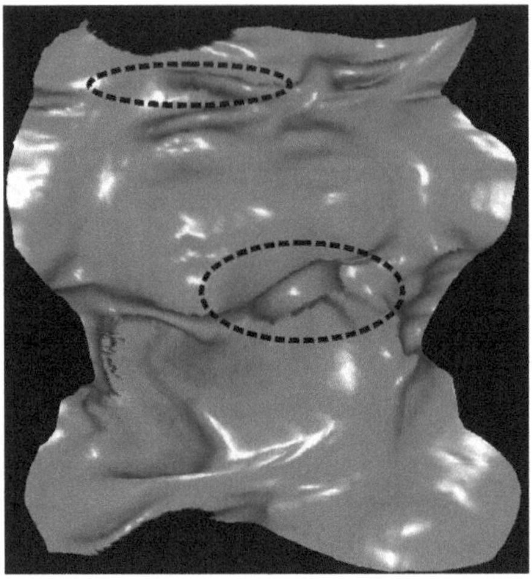

Fig. 7. Virtual colon unfolding using a topology change in the visualization [10] (colorplate on p. 220).

the data are often useful to increase the visibility of the most relevant objects. Examples demonstrate the significance of topology information for a broad spectrum of expressive focus+context visualizations.

6 Acknowledgments

The authors thank Tino Weinkauf and Holger Theisel for providing a polygonal model of the topological information of the flow in Figure 1. The work presented in this publication is carried out as part of the ex**vis**ation project (www.cg.tuwien.ac.at/research/vis/exvisation) supported by the Austrian Science Fund (FWF) under grant no. P18322.

References

1. S. Grimm, S. Bruckner, A. Kanitsar, and E. Gröller. Flexible direct multi-volume rendering in interactive scenes. In *Proceedings of Vision, Modeling, and Visualization'04*, pages 379–386, 2004.
2. H. Hauser. *Scientific Visualization: The Visual Extraction of Knowledge from Data*, chapter Generalizing Focus+Context Visualization, pages 305–327. Springer-Verlag, 2005.
3. J. Hladůvka. *Derivatives and Eigensystems for Volume-Data Analysis and Visualization.* PhD thesis, Vienna University of Technology, Austria, 2001.
4. H. Löffelmann. *Visualizing Local Properties and Characteristic Structures of Dynamical Systems.* PhD thesis, Vienna University of Technology, Austria, 1998.

5. Y. Sato, C. Westin, A. Bhalerao, S. Nakajima, N. Shiraga, S. Yoshida, and R. Kikinis. Tissue classification based on 3d local intensity structures for volume rendering. *IEEE Transactions on Visualization and Computer Graphics*, 6(2):160–180, 2000.
6. H. Theisel, T. Weinkauf, H.-C. Hege, and H.-P. Seidel. Saddle connectors - an approach to visualizing the topological skeleton of complex 3D vector fields. In *Proceedings of IEEE Visualization 2003*, pages 225–232, 2003.
7. H. Theisel, T. Weinkauf, H.-C. Hege, and H.-P. Seidel. Topological methods for 2d time-dependent vector fields based on stream lines and path lines. *IEEE Transactions on Visualization and Computer Graphics*, 11(4):383–394, 2005.
8. X. Tricoche, C. Garth, G. Kindlmann, E. Deines, G. Scheuermann, M. Ruetten, and C. Hansen. Visualization of intricate flow structures for vortex breakdown analysis. In *Proceedings of IEEE Visualization 2004*, pages 187–194, 2004.
9. J. J. van Wijk. Spot noise: Texture synthesis for data visualization. *Computer Graphics*, 25(4):319–318, 1991.
10. A. Vilanova. *Visualization Techniques for Virtual Endoscopy.* PhD thesis, Vienna University of Technology, Austria, 2001.
11. I. Viola. *Importance-Driven Expressive Visualization.* PhD thesis, Vienna University of Technology, Austria, 2005.
12. I. Viola, A. Kanitsar, and M. E. Gröller. Importance-driven feature enhancement in volume visualization. *IEEE Transactions on Visualization and Computer Graphics*, 11(4):408–418, 2005.
13. T. Weinkauf, H. Theisel, H.-C. Hege, and H.-P. Seidel. Topological construction and visualization of higher order 3D vector fields. In *Proceedings of Eurographics 2004*, pages 469–478, 2004.
14. X. Zheng and A. Pang. Topological lines in 3D tensor fields. In *Proceedings of IEEE Visualization 2004*, pages 313–320, 2004.

N-dimensional Data-Dependent Reconstruction Using Topological Changes

Zsolt Tóth[1], Ivan Viola[2], Andrej Ferko[1], and Eduard Gröller[3]

[1] Faculty of Mathematics, Physics, and Informatics, Comenius University, Bratislava, Slovakia {toth|ferko}@sccg.sk
[2] Department of Informatics, University of Bergen, Norway ivan@ii.uib.no
[3] Institute of Computer Graphics and Algorithms, Vienna University of Technology, Austria meister@cg.tuwien.ac.at

Summary. We introduce a new concept for a geometrically based feature preserving reconstruction technique of n-dimensional scattered data. Our goal is to generate an n-dimensional triangulation, which preserves the high frequency regions via local topology changes. It is the generalization of a 2D reconstruction approach based on data-dependent triangulation and Lawson's optimization procedure. The definition of the mathematic optimum of the reconstruction is given. We discuss an original cost function and a generalization of known functions for the n-dimensional case.

1 Introduction

The continuous reconstruction from discretely sampled data is an important part of data processing. Reconstruction is necessary to determine values at arbitrary positions, not just where the data sample is available. Such discrete data sets can be acquired by digital photography in 2D or computed tomography (CT) and magnetic resonance imaging (MRI) in 3D. Another way of data acquisition is mathematical simulation of certain phenomena used, e.g., in finite element methods.

In this work we introduce a reconstruction technique based on topological changes of triangulations. The topology of the resulting mesh is driven by the features represented in the data. Our optimization process turns an arbitrary triangulation of discretely sampled data into a feature-preserving mesh.

The most common reconstruction technique for regularly placed data is convolution based resampling using reconstruction filters. The drawback is that the features having other directions than the directions of the applied 1D basis functions are not reconstructed very well. This results in blurry artifacts at the border between different features. Geometrically based reconstruction, denoted as data-dependent triangulation (DDT) was introduced by Dyn et al. [6]. In Fig. 1 the output from convolution and triangulation based techniques is shown, at 1000% magnification. This kind of triangulation allows the generation of inevitable long and tiny triangles to preserve high-frequency features, and can be applied for any distribution of scattered data. The original DDT is limited to two dimensions and its extension to higher dimensions is not trivial. Its generalization into arbitrary dimension is the main scope of our work. In volume graphics this method can be applied for reconstruction using both regular and irregular grids. Therefore the target applications are tasks, where the reconstruction of sharp features is crucial. In this work we show the reconstruction results of this technique on regularly placed samples.

Fig. 1. Reconstruction result with bicubic filtering (left), data-dependent triangulation (right) at 1000% magnification.

The contributions of this paper are the following:

- n-dimensional reconstruction using a data-dependent triangulation approach
- a mathematical definition of the optimal reconstruction using triangulations
- a new cost function and generalization of existing functions.

This paper is structured as follows. In Sect. 2 we briefly survey previous work on DDT. Section 3 defines basic concepts of n-dimensional triangulations and the problem of the extension to higher dimensions is highlighted. In the same section the n-dimensional DDT is introduced. Section 4 presents a feature-preserving triangulation-algorithm and estimates the usability of the approach. In Sect. 5 we show a number of examples that demonstrate the effectiveness of our concept. Finally, in Sect. 6 we draw conclusions from the results and we outline future work possibilities.

2 Related Work

A general notion for interpolating given scalar data over an n-dimensional domain is called *scattered data interpolation*, and mathematically can be expressed as:

$$F(x_{i,1}, x_{i,2}, \dots, x_{i,n}) = z_i \,, \quad i = 1, \dots, m \,,$$

where $V = \{V_i = (x_{i,1}, x_{i,2}, \dots, x_{i,n}) \in \mathbf{R}^n\}$ is a set of not all cohyperplanar, distinct points in the domain space, and z_i are the measured data values. The goal of the reconstruction is to evaluate the function value $F(\boldsymbol{x})$ for an arbitrary point of the domain. In the rest of this section we give a short review of the related research work done in this field.

Data-dependent triangulation (DDT) is a geometrically based reconstruction technique developed by Dyn et al. [6]. It is a special case of optimal triangulations. A general survey on optimal triangulations was done by Bern and Eppstein [2]. DDT fits the measured data values with a triangle mesh, creating a piecewise linear interpolation. In contrary to other mesh generation methods, it maps the alignment of edges to the underlying data and organizes the simplicial structure into a feature-preserving mesh. The quality of the resulting triangulation is defined through a special function called *cost function* and the optimization process. Several algorithms have been developed and different optimization techniques have been applied to generate DDT in the 2D case. A genetic optimization technique was introduced by Kolingerová [10]. The DDT technique combined with simulated annealing was first introduced by Schumaker [18]. Typically, the above mentioned approaches assign the cost function values to the edges in the triangulations. Brown [4] came up with cost assigning to vertices. This idea is called vertex-based DDT, and it is a useful improvement of the basic approach.

The application of the DDT to image reconstruction has been done by Yu et al. [22]. Improving the image reconstruction quality has been the scope of our previous work [21]. Kreylos et al. [11] used DDT for image compression with a mesh decimation process, based on a simulated annealing optimization technique. A real-time version, limited to regularly placed image data was presented by Su and Willis [20]. The results from this simplified version are not as convincing as from other triangulation-based techniques. Battiato et al. [1] used the concept of a triangulation-based technique for creating vector format images from raster data.

Feature-preserving triangulations for volume data sets with a mesh refinement technique was studied by Marchesin et al. [14] and Roxborough and Nielson [17]. Both methods are based on the longest edge split approach. Other reconstruction techniques are numerical methods applicable for scattered data reconstruction, like C^1 methods in the 3D case by Nielson [16]. These techniques are not as good as a polygonal representation, which has better visual structuring and easier handling when compared to an analytical description.

An initial approach to extend DDT to 3D has been proposed by Lee [13]. Lee used the simulated annealing technique to get an optimal data-dependent mesh. This may result in artifacts, as we will show later in Sect. 5. In the literature, there is no description of the n-dimensional case. In the following section we introduce the general approach to n-dimensional triangulation.

3 *N*-dimensional Triangulation

A correct description of triangulation in arbitrary dimension requires to use some definitions from simplex theory. We selected Edelsbrunner‘s and Shah‘s terminology [7].

The convex hull (*conv*) of $k+1$ affinely independent points in $\mathbf{R}^n$, marked as set T, is a *k-simplex*, denoted by σ_T, where $0 \leq k \leq n$. For subsets $U \subseteq T$, simplices σ_U are called *faces* of σ_T. A collection of simplices, $\mathcal{K}$, is a *simplicial complex* if:

(i) The faces of every simplex in $\mathcal{K}$ are also in $\mathcal{K}$.
(ii) If σ_T, $\sigma_{T'} \in \mathcal{K}$, then $\sigma_T \cap \sigma_{T'} = \sigma_{T \cap T'}$.

Let V be a finite point set in $\mathbf{R}^n$. Usually a triangulation of V is defined as a simplicial complex so that V is the set of 0-simplices (vertices) and the underlying space of the complex is the convex hull of V. A simplicial complex $\mathcal{K}$ is a *triangulation* of V if:

(i) Each vertex of $\mathcal{K}$ is a point in V.
(ii) The underlying space of $\mathcal{K}$ is $conv(V)$.

The *content* of the n-simplex is its generalized volume (in 2D area, in 3D volume, etc.).

From the definition of the triangulation above we can see that it is a C^0 continuous reconstruction. The number of possible valid mesh configurations is exponential in the number of elements in the discrete data set.

Traditional (not data-dependent) triangulations usually avoid the generation of long, bad shaped k-simplices (triangles), because these simplices are not well suited for finite element methods used in simulations. However, such simplices are very suitable for reconstruction of areas where a function has high derivatives in one direction as compared to other directions. In images (2D domain case) such areas correspond to edges, in volume data (3D domain case) to surfaces of volumetric features. Our goal is to reconstruct feature boundaries sharply and to correctly use a reconstruction grid which adapts to the underlying data structures.

In Fig. 2 a part of an image with a boundary between two constant regions is shown. The reconstruction process first converts the image into a height-field representation based on the underlying sample (pixel) values. The height represents the intensity value at a particular pixel position. The initial triangulation is iteratively optimized in order to preserve the feature boundaries.

The differences to the initial mesh appear only close to the feature boundary. Figure 2 shows of the final height-field of the feature-preserving triangulation.

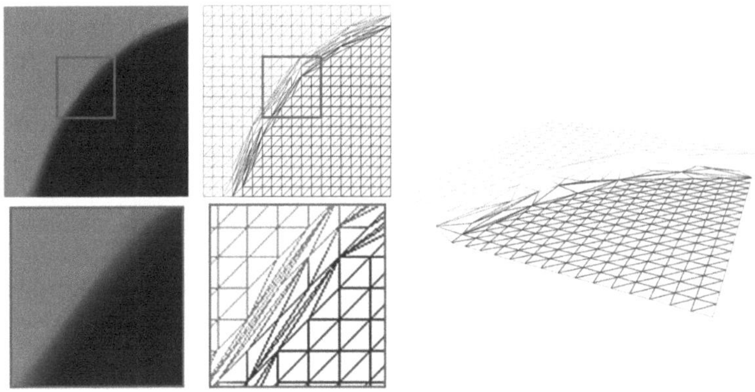

Fig. 2. Shape of the triangles at the border between different features.

In DDT applications cost functions are used to control the shape of the resulting mesh. The task of the optimization process is to improve the mesh via topological changes with regard to the cost function. Cost functions in 2D can be assigned to vertices (*vertex based* DDT), to edges, or to triangles. Most of the existing functions are defined for edges. Generally, cost functions assign a cost to every k-simplex in the triangulation, according to some local (not strictly geometrically based) property, where $k = 0, \ldots, n$ and n is the dimension of the domain space of the scattered data.

Different triangulations in 2D have the same number of edges. Thus, a topological change can not alter the number of edges in 2D. The goal of the reconstruction is to minimize the sum of the cost function weights of particular simplices. We get a structurally identical problem to the minimum-weight triangulation-problem [8], where the task is to find the triangulation with the minimum cost:

$$cost(\mathcal{K}_{optimal}) = min_{\mathcal{K} \in \Omega}(\textstyle\sum_{e \in \mathcal{K}}(cost(e)))$$

$\mathcal{K}_{optimal}$ is the triangulation generating the minimum sum among all possible configurations marked as Ω and e are simplices assigned with the cost function. The NP-hardness of this problem was shown by Mulzer and Rote [15]. Only its approximation can be computed in polynomial time. In previous work the global optimum for DDT was not defined and various heuristics were used for local improvements [11, 22].

A generalization to higher dimensions based on a constant number of simplices (edges in the 2D case) is not straightforward. The number of simplicial components of the n-dimensional triangulation in 3D and higher dimensions depends on the specific triangulation. For example if in 3D we decide to assign costs to faces (2-simplex), the sum changes not only due to the optimization steps but also depends on the number of faces in the triangulation (see Fig. 6). It is necessary to find a solution independent of mesh-connectivity changes. A vertex-based DDT satisfies this criterion, because the number of vertices (0-simplices) remains unchanged. However, designing a cost function for the vertex-based approach is not an intuitive task even in the 2D case. In higher dimensions this becomes even more difficult.

Our solution is based on the observation that the convex hull of the scattered points also remains unchanged. This means that the content is constant irrespective of the triangulation, and the task of reconstruction can be formulated as follows: the optimization process in the n-dimensional case should assign low weights to n-simplices which are well aligned with the underlying data. This means minimizing the weighted volumes of n-simplices summed over the entire space of the convex hull. The exact mathematical description is given as follows:

$$cost(\mathcal{K}_{optimal}) = min_{\mathcal{K}\in\Omega}(\textstyle\sum_{\sigma_n\in\mathcal{K}}(V(\sigma_n)\cdot w(\sigma_n)))$$

σ_n is an n-simplex, $V(\sigma_n)$ is its content, and the assigned weight function is $w(\sigma_n)$. $\mathcal{K}_{optimal}$ is the simplicial complex with minimal weighted volume in the set of possible configurations denoted as Ω. This observation is useful only if we can easily and intuitively find feature-preserving weight functions. A simple approach can use the variance property for weight assignment. Such an idea can be based on the preservation of low variance regions. Therefore the generation of $(n-1)$-simplices (faces) with low variance is preferred. This implies a weight function based on the variance of the function values in the n-simplex

$$w(\sigma_n) = Variance(z_{n_1}, z_{n_2}, \ldots, z_{n_{n+1}}),$$

where $w(\sigma_n)$ is the weight function for the σ_n n-simplex and z_{n_i}, $i = 1, 2, \ldots, n+1$ are the known function values in the simplex vertices. Figure 3 illustrates the example of a 3D domain case of a triangulation on a given surface. If the second type of triangulation will be chosen (see Fig. 6 image on the top right), the weighted volume value will be higher.

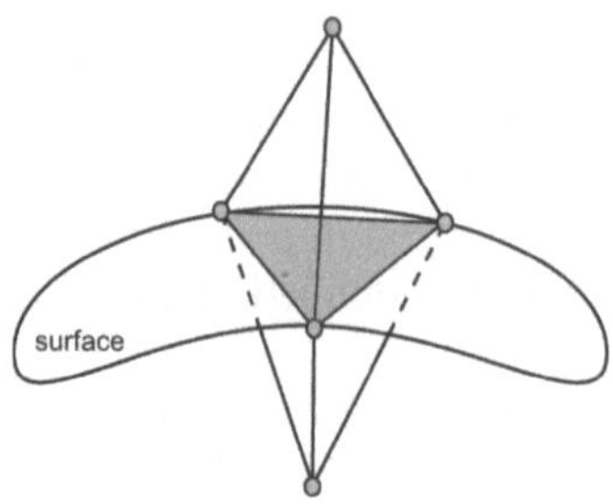

Fig. 3. Example of correct tetrahedralization according to the face variance.

Another useful property of triangulations in arbitrary dimensions is that $(n-1)$-simplices not lying at the boundary of the convex hull are exactly shared with two n-simplices. In 2D triangulations each interior edge is shared exactly by two triangles, in 3D each interior triangular face is shared exactly by two tetrahedrons, etc. This allows us to generalize most of the known cost functions from the 2D case for the $(n-1)$-simplices. Each n-simplex contains $(n+1)$ of $(n-1)$-simplices, for which the generalized feature-preserving cost can be evaluated. Averaging of these values gives a feature-fitting weight function for our technique. We illustrate this on a concrete cost function generalization to arbitrary dimensions. We have chosen a cost function from the 2D case with the most convincing result, i.e., Sederberg's cost function [22]. In Fig. 4 the cost function dependency for DDT is shown over a planar domain. T_1 and T_2 are triangles on the generated piecewise linear surface. They share a common edge, denoted as e. Sederberg's cost function is based on the angle α between the gradients of the planes containing triangles T_1, T_2, weighted by the magnitude of the gradients:

$$cost(e) = \| \bigtriangledown P_1\| \cdot \| \bigtriangledown P_2\| \cdot (1 - cos(\alpha)) = \| \bigtriangledown P_1\| \cdot \| \bigtriangledown P_2\| - \bigtriangledown P_1 \cdot \bigtriangledown P_2$$

$\bigtriangledown P_1$, $\bigtriangledown P_2$ are the gradients of the planes containing T_1, T_2, and $\|P_1\|$, $\|P_2\|$ are their magnitudes. Angle α is the angle between these gradients.

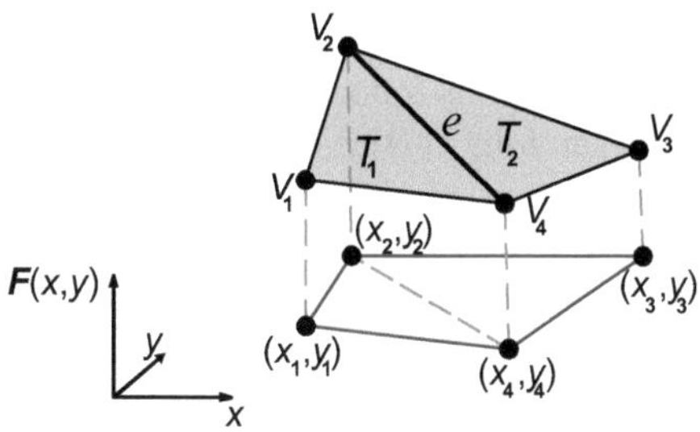

Fig. 4. Illustration of geometric dependencies in 2D data dependent triangulation.

Its straightforward generalization to higher dimensions looks as follows:

$$cost(\sigma_{n-1}) = \| \bigtriangledown P_1\| \cdot \| \bigtriangledown P_2\| - \bigtriangledown P_1 \cdot \bigtriangledown P_2\,,$$

where linear polynomials $P_1(\boldsymbol{x})$ and $P_2(\boldsymbol{x})$ represent the hyperplanes. These hyperplanes are calculated from the n-dimensional domain and the function values of the data function:

$$P_i(\boldsymbol{x}) = a_{i,1}x_1 + a_{i,2}x_2 + \ldots + a_{i,n}x_n + a_{i,n+1}\,, \quad i = \{1, 2\}\,.$$

$\bigtriangledown P_i$ is the gradient of the hyperplane, and $\| \bigtriangledown P_i\|$ is the magnitude of the gradient:

$$\bigtriangledown P_i = (a_{i,1}, a_{i,2}, \ldots, a_{i,n})\,, \quad i = \{1, 2\}\,,$$
$$\| \bigtriangledown P_i\| = \sqrt{a_{i,1}^2 + a_{i,2}^2 + \ldots + a_{i,n}^2}\,, \quad i = \{1, 2\}\,.$$

With the help of these generalized cost functions we can describe a feature-preserving mesh, as follows:

$$w(\sigma_n) = \frac{\sum_{\sigma_{n-1} \in \sigma_n} cost(\sigma_{n-1})}{(n+1)}$$

σ_n is an n-simplex and $w(\sigma_n)$ is its weight function. The generalized cost functions of the $(n-1)$-dimensional subsets (faces) of this simplex are averaged. The $(n-1)$-simplices are marked in the sum as σ_{n-1}.

Other geometrical properties can also be involved in the weight function design. On the basis of the shape of $(n-1)$-simplices we can consider their weighted sum instead of the unweighted sum. Having feature-fitting weight functions the task is to solve the construction of an optimized mesh. This problem is treated in the next section.

4 Content-Based Data-Dependent Reconstruction

As we mentioned earlier, the feature-preserving property of the reconstruction technique is achieved by specific quality-improving operations on the topology of the triangulation. In 2D this kind of topology transformation is called *edge flip* and it changes the topology as follows: In Fig. 4 the edge e can be replaced with the edge $V_1 - V_3$ if the four vertices form a strictly convex quadrilateral. Via flip improvements the cost of the triangulation is decreased iteratively, and we are getting a feature-preserving mesh. The 2D triangulation algorithm based on this idea is called *Lawson's optimization process* [12].

In higher dimensions the notion of an edge flip generalizes to *bistellar flips*. Bistellar flips include removals from and insertions into triangulations. We are interested only in such transformations where the number of vertices does not change. This general topological operation is based on Radon's theorem from convex geometry [7]. Each transformation can be interpreted as a projection of a simplex into a lower dimension. The views of the simplex from antipodal points of view in the direction of the projection introduces the two configurations, before and after the flip, as is illustrated in Fig. 5. For example if we project a tetrahedron into a plane we get an edge flip. Projections which cause degeneracies in the projected complex can be interpreted as bistellar flips of lower dimension in the simplex (of lower dimension) where the degeneracy occurred.

Let us describe the set of bistellar flips in 3D:

- A face can be changed into an edge as is illustrated in Fig. 6 (top part). This is called a 2–3 flip.
- An edge can be changed to a face as is illustrated in Fig. 6 (top part). This is called a 3–2 flip.

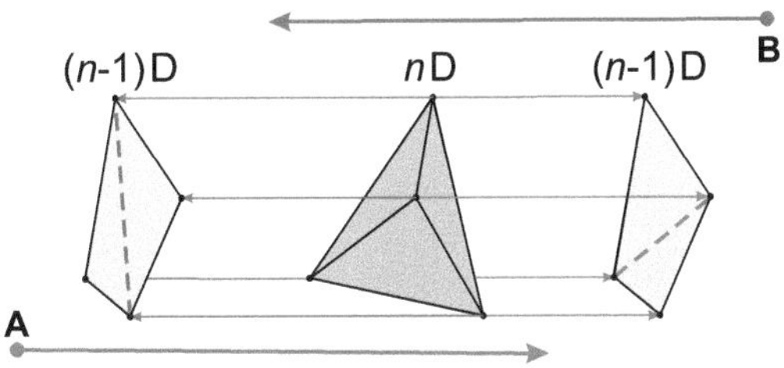

Fig. 5. Illustration of the bistellar flip. **A** and **B** are the antipodal points of view of the projection of an n-simplex (center) to a lower dimension (left, right).

- A degenerate case occurs when six vertices form four adjacent tetrahedrons with one common edge as is illustrated in Fig. 6 (bottom part). If four of these vertices are coplanar and the remaining vertices are divided by this plane, there are two possible ways how to tetrahedralize this structure. This operation is called 4–4 flip.

In higher dimensions the situation is the following:

- Let us consider an n-dimensional space, and a simplicial complex $\mathcal{M}$ of n-simplices which share a common edge. If $\mathcal{M}$ has $n+2$ vertices and forms a convex space then there exist exactly two triangulations of $\mathcal{M}$. One that includes the common edge and one that instead includes a hyperface formed by the vertices not related to the removed edge.
- All other possibilities are degeneracies and can be described as bistellar flips in lower dimension.

Unlike in the 2D case it is not proven that with these bistellar flips one can get from an arbitrary triangulation to any other possible configuration. To our knowledge it is proven only in the 3D case when triangulating a convex polytope [3].

Another possibility for mesh improving is to investigate more complex topological transformations, like the *edge-removal* operation in the 3D case. It is a transformation that removes a single edge from the mesh along with all tetrahedra that contain it. This operation can be composed from a series of bistellar flips, but in that way the optimalization can get stuck in local optima. An effective implementation was discussed in the work of Shewchuk [19].

To make our concept more general we define the set $\mathcal{T}_{\mathcal{K}}$ of topological transformations of triangulation $\mathcal{K}$. The members of this set can be selected arbitrarily. Reconstruction will be done considering the members of this set. Good reconstruction results can be expected from a set of $\mathcal{T}_{\mathcal{K}}$, that generates all possible topological transformations at the given dimension. We are interested in those topological operations, which remove a specific k-simplex $k = 1, \ldots, n-1$ and change the topology of n-simplices containing the removed k-simplex.

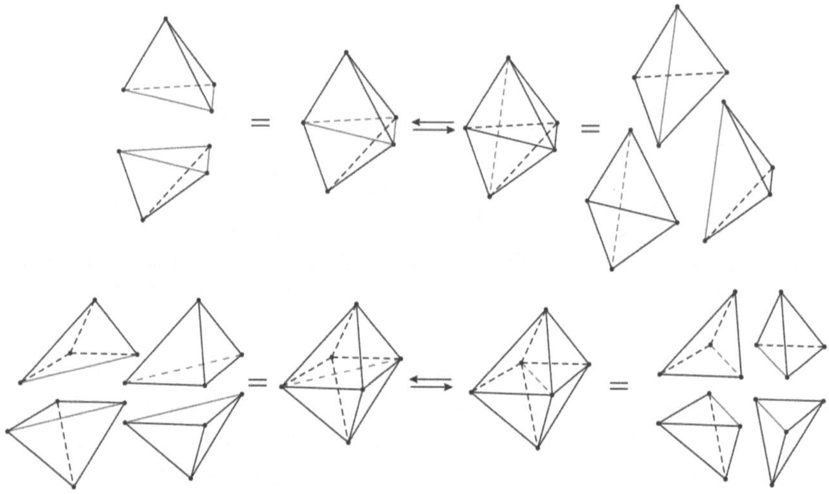

Fig. 6. Illustration of the 2–3 flip and 3–2 flip topological transformations (top image) and the 4–4 flip topological transformation (bottom image).

Let us have a triangulation $\mathcal{K}$ for the n-dimensional domain. The k-simplex $k = 1, \ldots, (n-1)$ is *locally optimal* with regard to $\mathcal{T}_{\mathcal{K}}$ and to a given cost function if:

- the k-simplex cannot be removed from the triangulation by applying one of the topological transformations from $\mathcal{T}_{\mathcal{K}}$
- there is a topological transformation in $\mathcal{T}_{\mathcal{K}}$ that removes the k-simplex (let us denote this k-simplex as *flippable*), but the triangulation cost does not decrease after the applied changes.

Triangulation $\mathcal{K}$ is *locally optimal* if topological transformations from $\mathcal{T}_{\mathcal{K}}$ cannot improve its weighted content.

We introduce the generalization of Lawson's algorithm which creates locally optimal DDT. The pseudocode of the algorithm is presented in Fig. 7. At first an initial triangulation is generated, as it is written in *line*2. We suggest for this purpose the Delaunay triangulation. The initial triangulation should connect each vertex with its closest neighbors. Due to the duality with Voronoi diagrams, the Delaunay triangulation satisfies this criterion. Creation of good initial triangulations speeds up the running time, as the algorithm needs less iterations to converge to a locally optimal solution. For image and volume data-reconstruction this means the following: The optimization process will leave the initial triangulation unchanged in low frequency areas, only in high frequency areas it will generate long feature-preserving simplices.

After creating the initial triangulation its weighted content is evaluated in *line*3. The *oldCost* variable is used to store the triangulation cost of the previous iteration step. Two lists are initialized in *line*5 and *line*6. $List_{active}$

contains all the simplices whose local optimality is tested in the given iteration step. In the beginning this list contains all $1, \ldots, n-1$ simplices of the triangulation $\mathcal{K}$. The second list $List_{candidate}$ is the container of simplices whose local optimality could change due to the applied transformations from $\mathcal{T}_{\mathcal{K}}$. We denote these simplices with σ_k. At the beginning of the optimization process $List_{candidate}$ is set to empty. In a given iteration local optimality of the members of $List_{active}$ are tested. If there exists a cost-reducing topological transformation from $\mathcal{T}_{\mathcal{K}}$ then it is applied. The tested simplex is removed from $List_{active}$. Into $List_{candidate}$ those simplices are added whose local optimality could change. At the end of each iteration step $List_{active}$ is empty and $List_{candidate}$ contains all the simplices that will be tested in the next iteration step. The algorithm stops, if the triangulation cost did not improve in the last iteration step. This procedure results in a locally optimal triangulation $\mathcal{K}$ with regard to a set of topological transformations $\mathcal{T}_{\mathcal{K}}$ and a given cost function.

It is evident that the described algorithm stops after a final number of iteration steps. In each step we decrease the overall cost of the triangulation and the number of possible triangulations is finite. The described idea of a weighted content-based DDT can be used with other optimization techniques also. It is possible to construct a simulated annealing, look-ahead, or genetic optimization approach based on this idea.

5 Experiments and Results

DDT triangulations can be computed by stochastic processes. Such a technique is the simulated annealing optimization approach. Its usage can improve the approximation level of the algorithm to achieve results which are closer to the global optimum than the described generalized Lawson's optimization process. However, according to our experience simulated annealing does not create convincing results. The initial stages of simulated annealing apply topological transformations on the mesh which increase the cost of the triangulation. This results in long, bad shaped simplices in areas without high-frequency features. Our goal was to create these bad shaped simplices only at high frequency areas. In Fig. 8 a comparison between results from Lawson's optimization process and the simulated annealing technique is depicted, both using Sederberg's cost function. The reconstruction is displayed at 600% magnification. The mesh of the resulting triangulation and the reconstructed 2.5D terrain are also shown. In low frequency areas (e.g., cheek area below the eye) simulated annealing creates long and tiny triangles. Their generation is in this case undesired. From the mathematical point of view this observation for image and volume data reconstruction can be formulated as follows: A high-quality reconstruction should give a good approximation of the minimum-weight triangulation, but should not change the topology in low-frequency regions.

Generalized Lawson's optimization process

Input: scattered data
Output: locally optimal triangulation

```
begin
  create the initial triangulation K;
  Cost = cost(K);
  oldCost = Cost + 1;
  List_active = {∀σ_k ∈ K,  k = 1, ..., n − 1};
  List_candidate = ∅;
  while (Cost < oldCost)
  {
    for (List_active members)
    {
       if (not locally optimal)
       {
         apply transformation from T_K;
         remove from List_active;
         for (σ_k whose local optimality could be changed)
         {
           if (σ_k is not member of List_active and List_candidate)
              add σ_k to List_candidate;
         }
       }
       else
         remove σ_k from List_active;
    }
    oldCost = Cost;
    Cost = cost(K);
    List_active = List_candidate;
    List_candidate = ∅;
  }
  locally optimal K created;
end
```

Fig. 7. The pseudocode of the n-dimensional DDT algorithm.

In Fig. 9 a comparison between a convolution based approach – trilinear interpolation, and DDT in 3D – using the generalized Lawson's optimization with bistellar flips and the variance-based weight function, is shown. The artificial test data set was a sphere embedded into a cube at $32 \times 32 \times 32$ resolution. One can see that the surface of the sphere is reconstructed better by the DDT approach.

In Fig. 10, we have reconstructed the hydrogen dataset of resolution $32 \times 32 \times 32$. In the left image trilinear reconstruction has been used which resulted in clearly noticeable staircase artifacts. These artifacts are eliminated

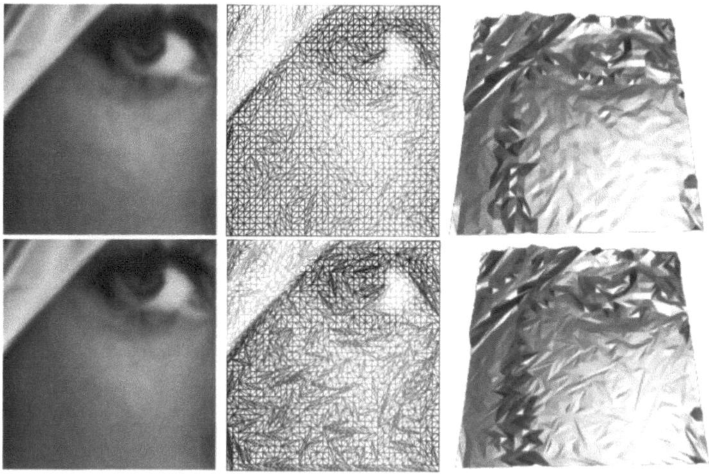

Fig. 8. Results from Lawson's optimization process (upper row), and from simulated annealing (lower row) at 600% magnification. The last column shows the triangulation of the height field.

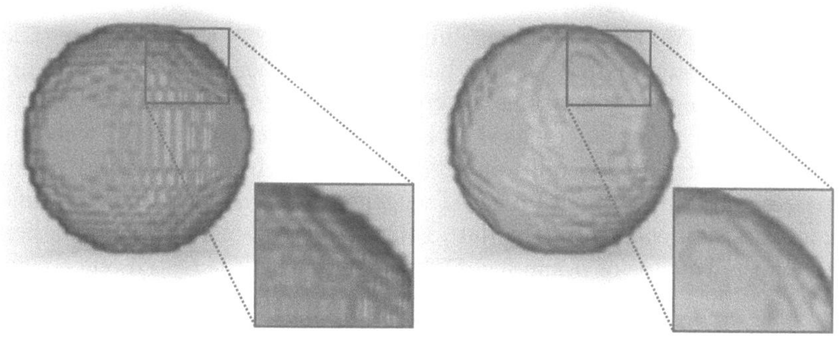

Fig. 9. Results from trilinear interpolation (left), and from the generalized Lawson's optimization DDT approach with the variance-based weight function and bistellar flips (right). (Colorplate on p. 221.)

in the right image, where the DDT-based reconstruction scheme has been applied. This observation indicates the superiority of feature-preserving DDT reconstruction over traditional trilinear interpolation.

The implementation of the 3D DDT was done with the CGAL 3.1 library [5]. For the rendering of the datasets the VTK 5.0 library [9] was used. This software tool works with both structured (rectilinear) and unstructured

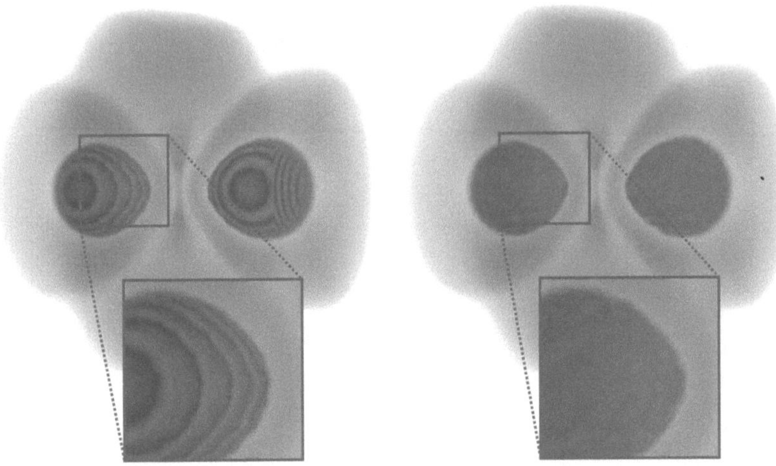

Fig. 10. Hydrogen dataset: The left image shows the result from trilinear interpolation suffering from noticeable artifacts. The right one shows improved results achieved by DDT-based reconstruction with the bistellar flips and the variance-based weight function (colorplate on p. 222).

grids. The possible differences caused by the usage of different renderers are minimized in this way.

In Sect. 3 two types of weight functions have been described: the variance-based approach and an approach based on the generalization of existing cost functions to higher dimensions. There is a significant difference between these techniques. The variance-based weight function is clearly a C^0 continuous approach and depends only on the data values from the tested n-simplex vertices. The weight function based on generalized cost functions has different properties. It depends on the data values from the tested n-simplex vertices and on the vertices of n-simplices which share with the tested one a common $(n-1)$-simplex. For this reason these weight functions can be classified as *near* C^1 continuous weight functions. Because of this difference the running time of the DDT with the variance-based cost function is lower than the running time of the DDT with generalized cost functions. The reason is that the weight function evaluation is computationally cheaper, and the number of simplices whose local optimality could change is smaller.

6 Conclusions and Future Work

We have introduced a new method using topological changes for n-dimensional data-dependent triangulation. This enables a better visualization of the local topology of sharp features both for regular and sparse grids in arbitrary dimensions. The topology of the reconstruction grid is driven by the topology of the underlying data. The potential of this new reconstruction technique results in many future work directions:

Our feature preserving triangulation can be used for data compression. If we assign a cost to vertices we have information about how important the vertices are. With mesh decimation techniques the non important vertices can be removed.

A straightforward use of the described technique for the reconstruction of time-varying volume data is possible. Here reconstruction errors from known image processing methods are more disturbing than in a static reconstruction.

Optimal triangulations in 2D are relatively well explored. Higher dimensional optimal triangulations represent a novel area for investigation. We hope that our contribution improves the theory in this research area and its application provides a new tool for practical scattered data reconstruction.

7 Acknowledgments

Funds for the support of this research have been allocated by the Rectorate of the Comenius University, VEGA grant no. 1/3083/06, and the Ausseninstitut der Technischen Universität Wien. Part of the work presented in this publication is carried out within the PVG project (`http://www.cg.tuwien.ac.at/research/vis/pvg/`) funded by the Austrian Science Fund (FWF) grant no. P18547N13 and from a Grant of Comenius University no. UK/362/2006.

References

1. L. Battiato, G. Gallo, and G. Messina. SVG rendering of real images using data dependent triangulation. In *Proceedings of Spring Conference on Computer Graphics 2004*, pages 191–198, 2004.
2. M. Bern and D. Eppstein. Mesh generation and optimal triangulation. In *Computing in Euclidean Geometry*, number 4, pages 47–123. World Scientific, second edition, 1995.
3. S. N. Bespamyatnikh. Enumerating triangulations of convex polytopes. In *Proc. Conf. Discrete Models: Combinatorics, Computation, & Geometry (DM-CCG 2001)*, number AA.
4. J. Brown. Vertex based data dependent triangulations. *Computer Aided Geometric Design*, 8(3):239–251, 1991.
5. CGAL, http://cgal.org/. *Documentation*, Computational Geometry Algorithms Library 3.1 edition, 2004.
6. N. Dyn, D. Levin, and S. Rippa. Data dependent triangulations for piecewise linear interpolation. *IMA Journal of Numerical Analysis*, (10):137–154, 1990.
7. H. Edelsbrunner and N. R. Shah. Incremental topological flipping works for regular triangulations. In *SCG '92: Proceedings of the 8th Annual Symposium on Computational Geometry*, pages 43–52. ACM Press, 1992.

8. A. Ferko, L. Niepel, and T. Plachetka. Criticism of hunting minimum weight triangulation edges. In *Proceedings of the 12th Spring Conference on Computer Graphics*, pages 259–264, 1996.
9. Kitware Inc., http://public.kitware.com/VTK/. *Documentation*, Visualization ToolKit 5.0 edition, 2005.
10. I. Kolingerová. Genetic approach to data dependent triangulations. In *Proceddings of Spring Conference on Computer Graphics 1999*, pages 229–238, 1999.
11. O. Kreylos and B. Hamann. On simulated annealing and the construction of linear spline approximations for scattered data. *IEEE Transactions on Visualization and Computer Graphics*, 7(1):17–31, 2001.
12. C. L. Lawson. Software for C^0 interpolation. *Mathematical Software III.*, pages 161–194, 1977.
13. K. Lee. Three-dimensional medical image modeling of scattered data based on data-dependent criteria. In *Proc. SPIE Vol. 4117, Vision Geometry IX*, pages 91–99, 2000.
14. S. Marchesin, J. M. Dischler, and C. Mongenet. 3D ROAM for scalable volume visualization. In *2004 IEEE Symposium on Volume Visualization and Graphics (VV'04)*, pages 79–86. IEEE Computer Society, 2004.
15. W. Mulzer and G. Rote. Minimum-weight triangulation is NP-hard. In *Proc. 22nd Ann. Symp. on Computational Geometry (SoCG06, Sedona, USA)*. ACM Press, to appear, 2006.
16. G. M. Nielson. Scattered data modeling. *IEEE Comput. Graph. Appl.*, 13(1):60–70, 1993.
17. T. Roxborough and G. M. Nielson. Tetrahedron based, least squares, progressive volume models with application to freehand ultrasound data. In *Proceedings of the 11th IEEE Visualization 2000 Conference (VIS 2000)*, pages 93–100. IEEE Computer Society, 2000.
18. L. L. Schumaker. Computing optimal triangulations using simulated annealing. In *Selected Papers of the International Symposium on Free-form Curves and Free-form Surfaces*, pages 329–345. Elsevier Science Publishers B. V., 1993.
19. J. R. Shewchuk. Two discrete optimization algorithms for the topological improvement of tetrahedral meshes. Unpublished manuscript. *Computer Science Division, University of California at Berkeley*, 2002.
20. D. Su and P. Willis. Image interpolation by pixel-level data-dependent triangulation. *Comput. Graph. Forum*, 23(2):189–202, 2004.
21. Z. Toth. Towards an optimal texture reconstruction. In *CESCG 2000 - 2005 Best Paper Selection*, pages 197–212. 2006.
22. X. Yu, B. S. Bryan, and T. W. Sederberg. Image reconstruction using data-dependent triangulation. *Computer Graphics and Applications*, 21(3):62–68, 2001.

A

Colorplates

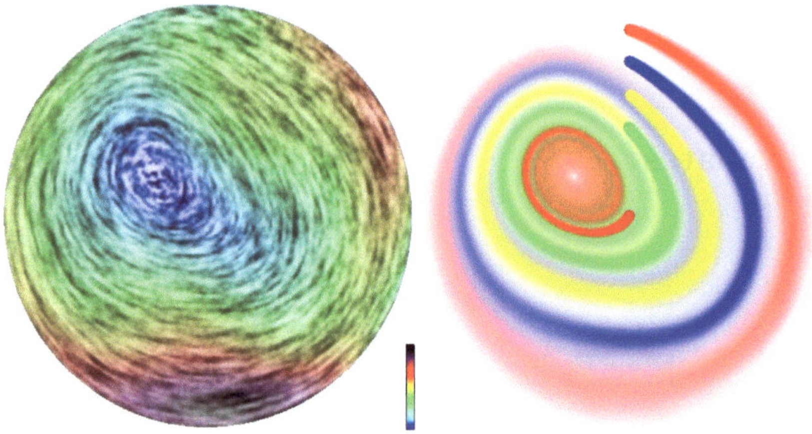

Fig. 1 (p. 2). Visualization of flow around a critical point using texture advection and dye injection [35]. In contrast to these methods, topology-based methods extract and visualize critical points directly.

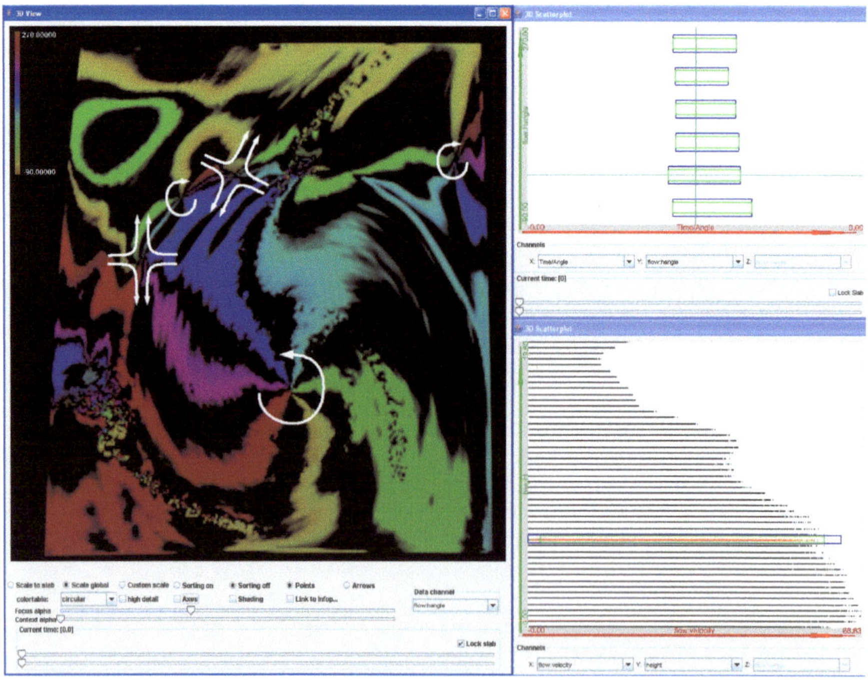

Fig. 3 (p. 6). Here, the topology of a hurricane is visualized and depicted with the SimVis system.

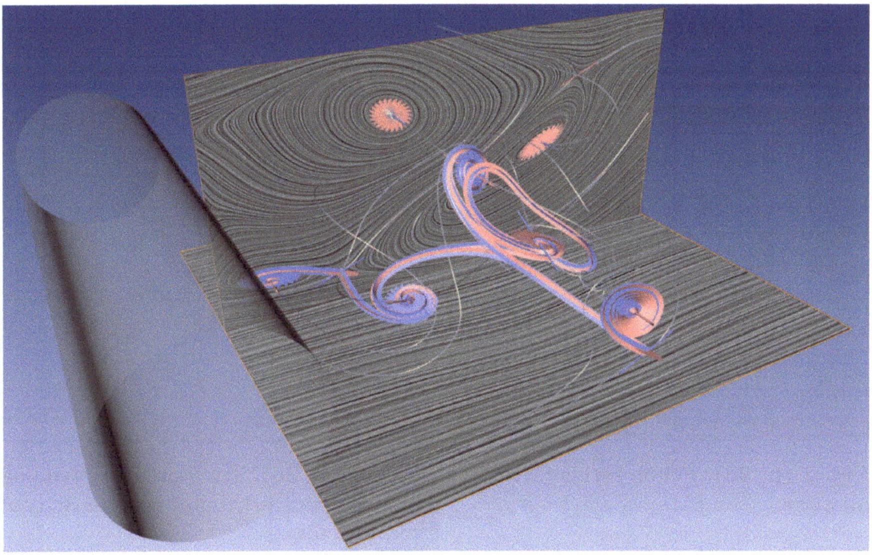

Fig. 4 (p. 10). Visualization of flow past a circular cylinder using critical points and saddle connectors. *Image courtesy of H. Theisel et al.*

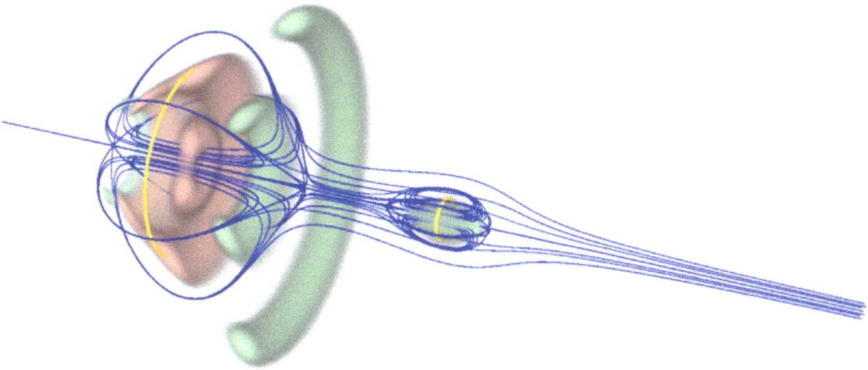

Fig. 5 (p. 11). The visualization of a vortex breakdown bubble. Flow topology is depicted with stagnation points in red, singularity paths in yellow and streamlines in blue. *Image courtesy of X. Tricoche et al.*

Fig. 2 (p. 28). Pelton turbine with five injectors.

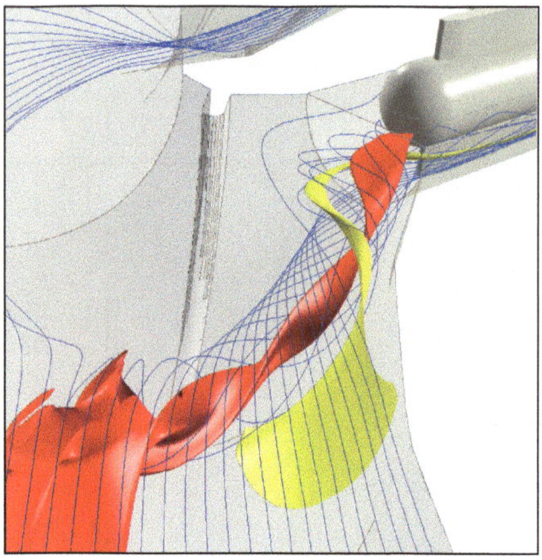

Fig. 3 (p. 28). Two vortices extending from the ring distributor into the first (of six) injectors.

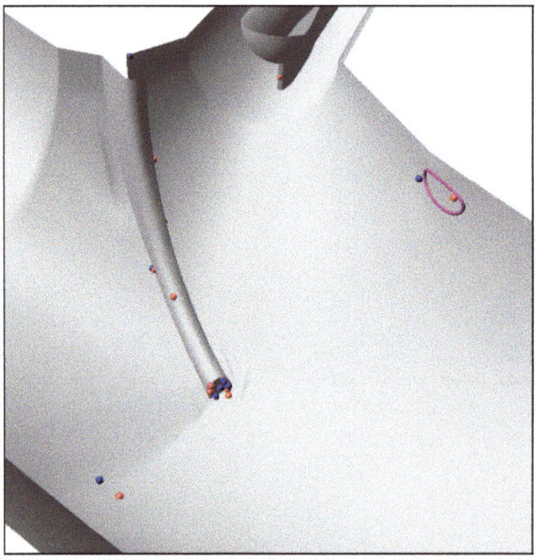

Fig. 4 (p. 28). Extracted interior (blue) and boundary (red) critical points. Periodic orbit (magenta).

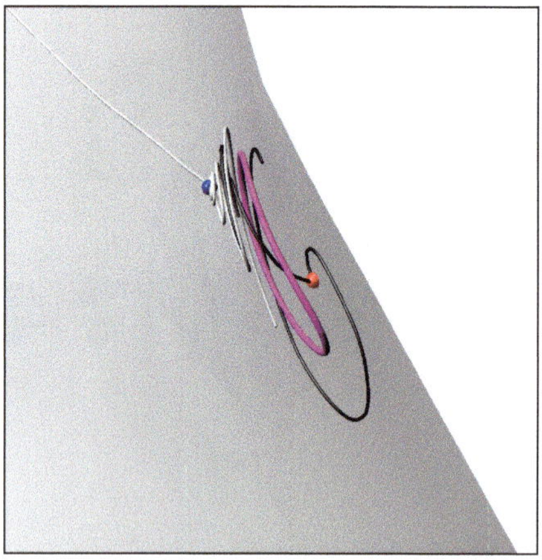

Fig. 5 (p. 28). Streamlines seeded near the boundary critical point (black) and the interior critical point (white).

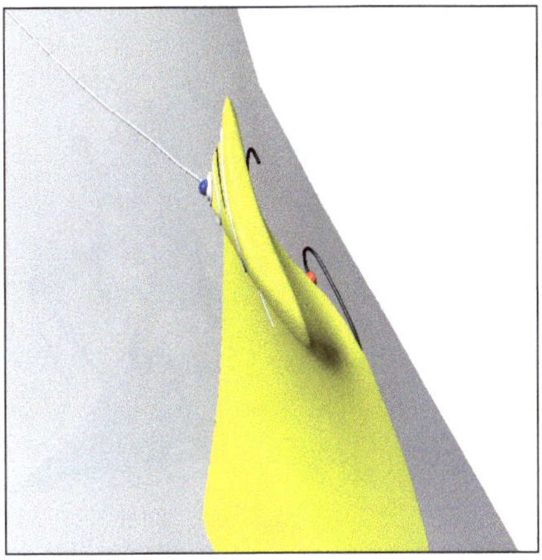

Fig. 6 (p. 29). Stable manifold of interior critical point.

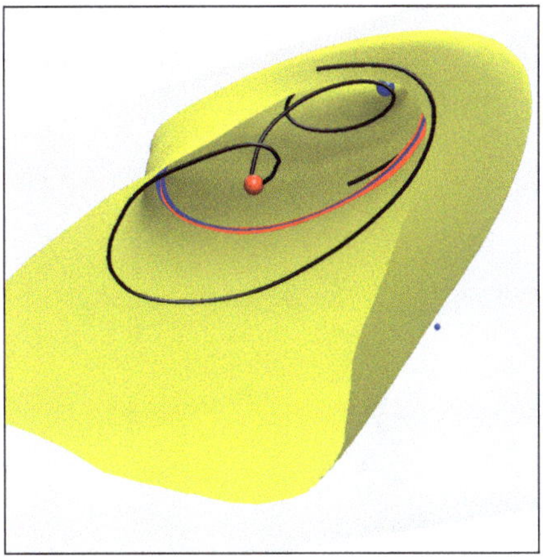

Fig. 7 (p. 29). View from the wall with stable and unstable manifolds of the periodic orbit (red and blue stream surfaces).

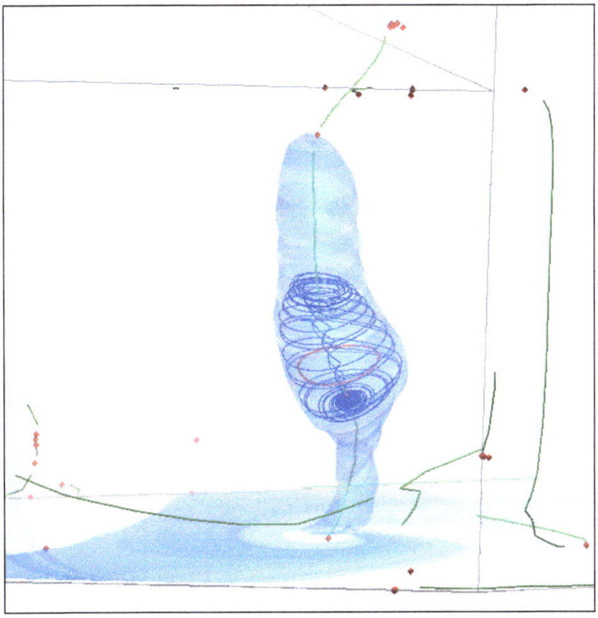

Fig. 8 (p. 30). Tornado-type separation and vortex in the draft tube dataset. Stream surface (transparent blue) starts at saddle and goes upstream enclosing a vortex breakdown bubble (blue streamline) containing a periodic orbit (red). Critical points (red) and vortex core lines (green).

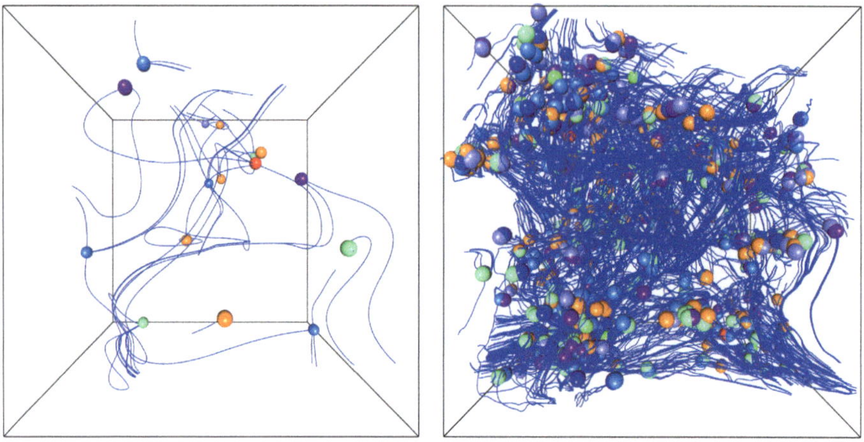

Fig. 1 (p. 45). A random generated test dataset. Critical points are color coded according to their classification. Foci are shown in red and green hues, while saddles are colored in blue tones. Streamlines are seeded in the vicinity of critical points. *Left:* The 16 critical points of the original data set. *Right:* After adding noise 1307 simple critical points can be detected.

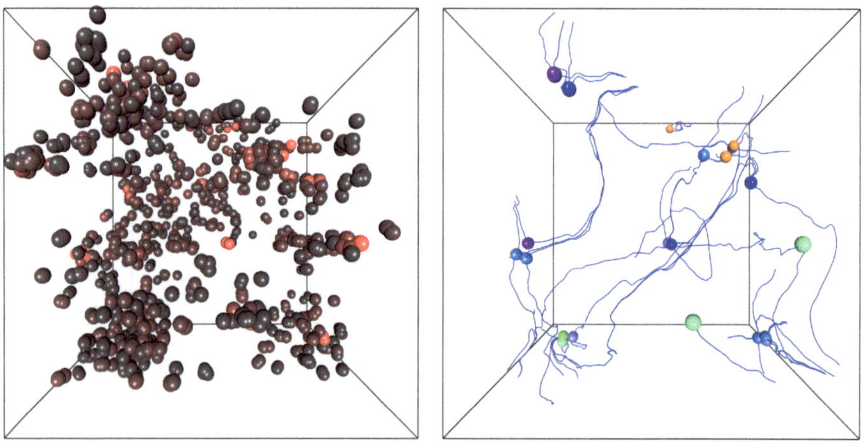

Fig. 2 (p. 46). The same test dataset as shown in Fig. 1. *Left:* Scale-space lifetime of the critical points in the interval $\tau = 0 \ldots 1.5$ computed by the scale-space tracking algorithm. Red color indicates stable critical points, while gray colored points are very short-lived. *Right:* Critical points filtered by their lifetime. Only points that persist at $\tau_{\min} = 1.0$ are shown and used to seed streamlines in their vicinity.

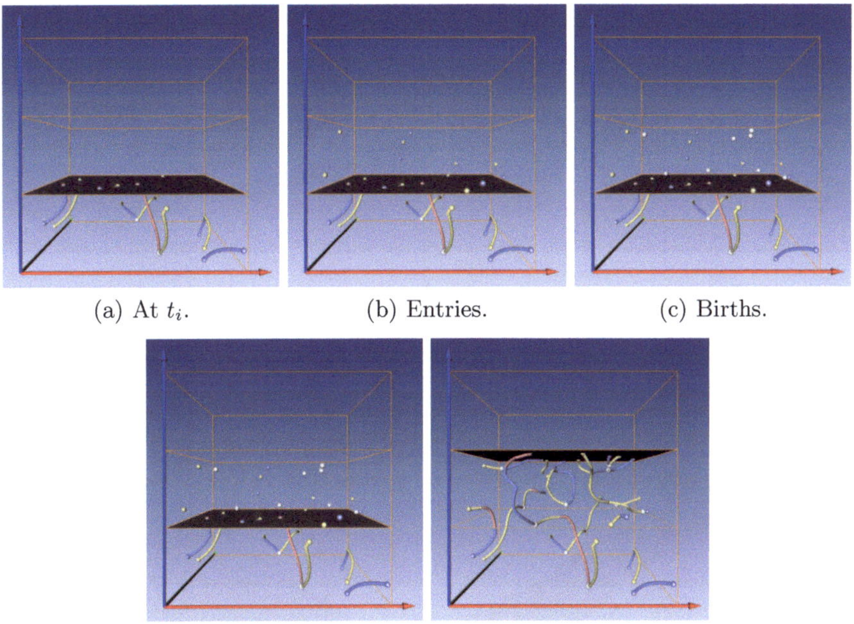

(a) At t_i. (b) Entries. (c) Births.

(d) Integration. (e) At t_{i+1}.

Fig. 7 (p. 61). Application of algorithm 3: critical points tracked in one sweep.

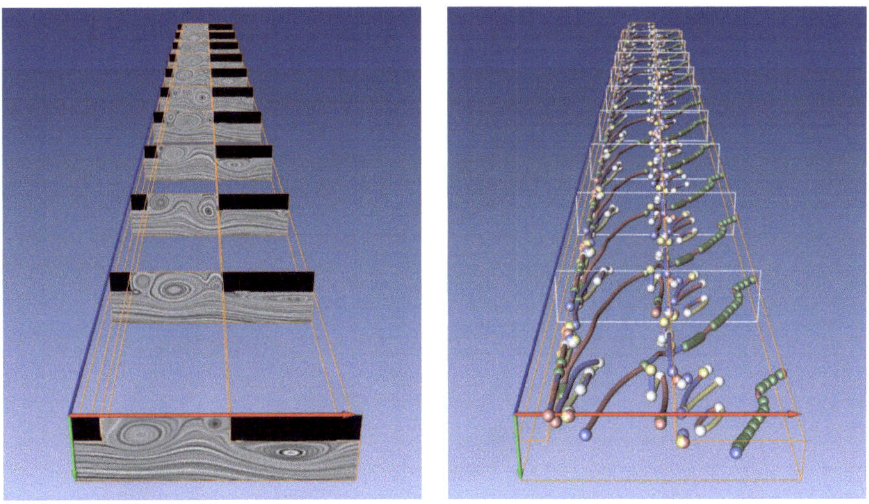

(a) LIC planes: 10 sections. (b) Tracked critical points.

Fig. 8 (p. 62). Cavity data set consisting of 1000 time steps. Algorithm 3 has been applied onto the 10 depicted sections consisting of 100 time steps each.

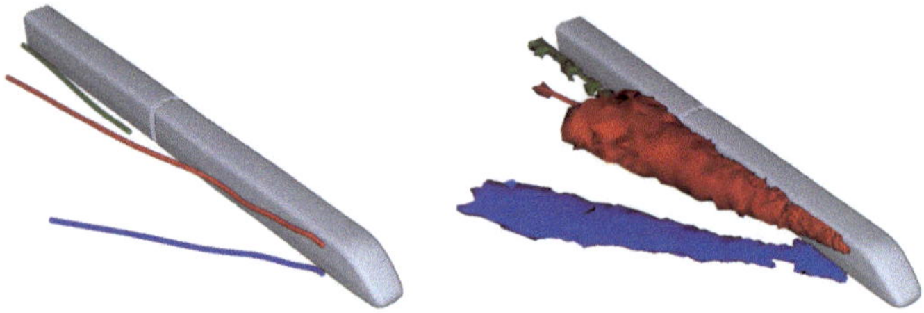

Fig. 3 (p. 74). The left picture shows the vortex core lines of the train-dataset. In the right picture the corresponding vortex core regions are depicted. (Note that the vortex core regions do not cover the vortex core lines completely. We assume that the vortices level off at the end.)

Fig. 4 (p. 75). Boundaries of characteristic sets of $\mathcal{G}_{Vortex}$ characterizing these parts of the flow that stream only in one vortex region (top down): $R \wedge \bar{G} \wedge \bar{B}$, $\bar{R} \wedge G \wedge \bar{B}$, $\bar{R} \wedge \bar{G} \wedge B$.

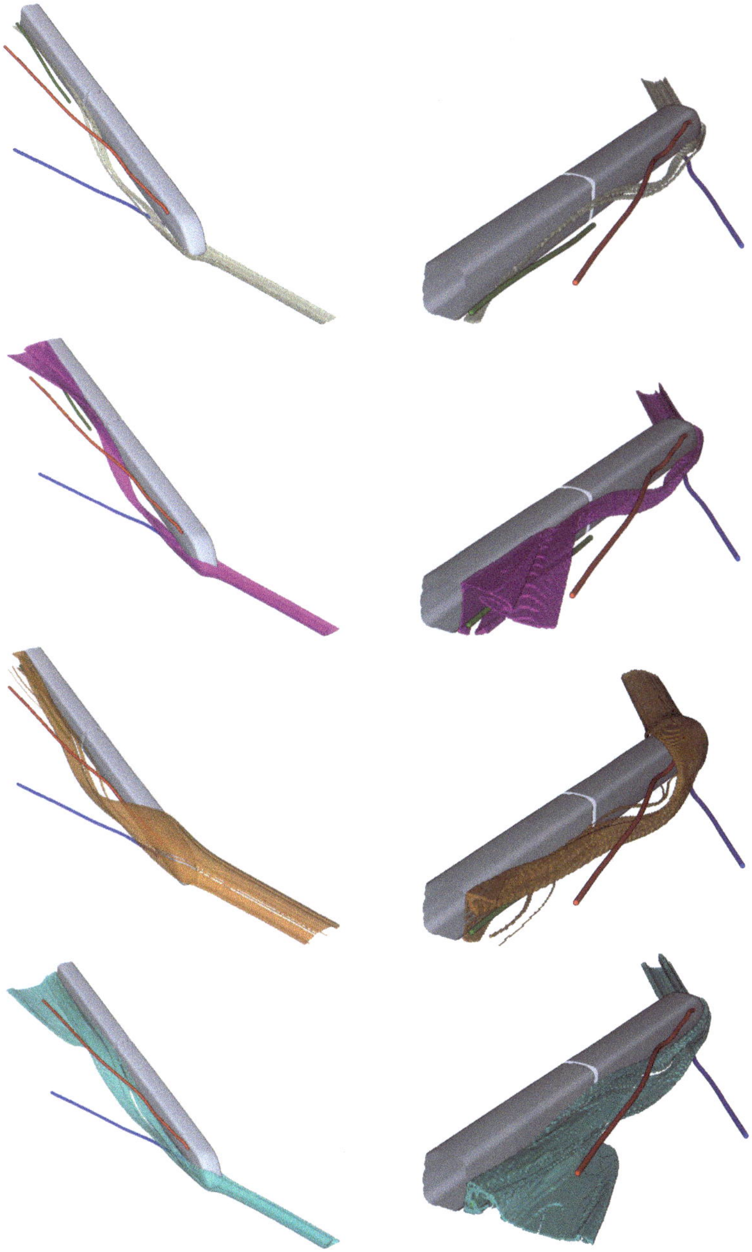

Fig. 5 (p. 76). Boundaries of characteristic sets of $\mathcal{G}_{Vortex}$ characterizing these parts of the flow that stream in more than one vortex region (top down): $R \wedge G \wedge B$, $R \wedge \bar{G} \wedge B$, $R \wedge G \wedge \bar{B}$, $\bar{R} \wedge G \wedge B$ ($\bar{R} \wedge \bar{G} \wedge \bar{B}$ (i.e. flowing in no vortex region) is not shown.) The second column shows the same characteristic sets from another view.

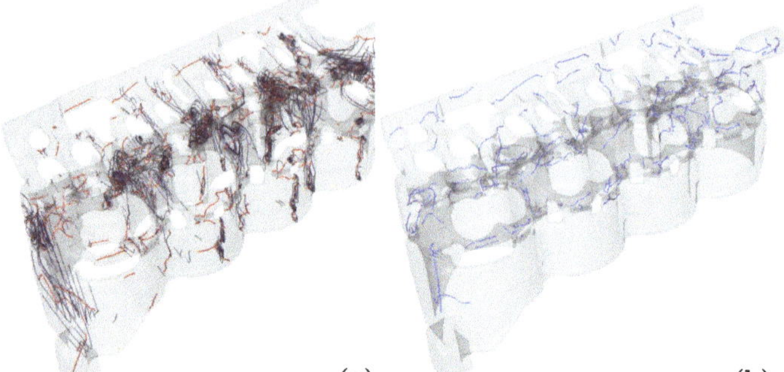

(a) (b)

Fig. 1 (p. 81). Cutting plane topology, revealing flow structures perpendicular to the dominant flow direction – streamlines are depicted in blue, longitudinal vortex cores are shown in red. (**b**) Vortex core line extraction using the method from Sujudi and Haimes.

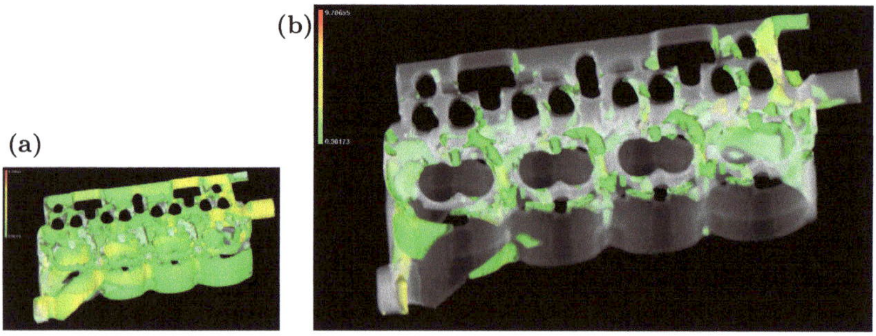

(a) (b)

Fig. 3 (p. 84). (**a**) The visualization of all regions of forward-longitudinal flow; color-mapping reflects velocity magnitude. (**b**) The result of selecting all regions of reverse-longitudinal flow, the inverse of the left selection.

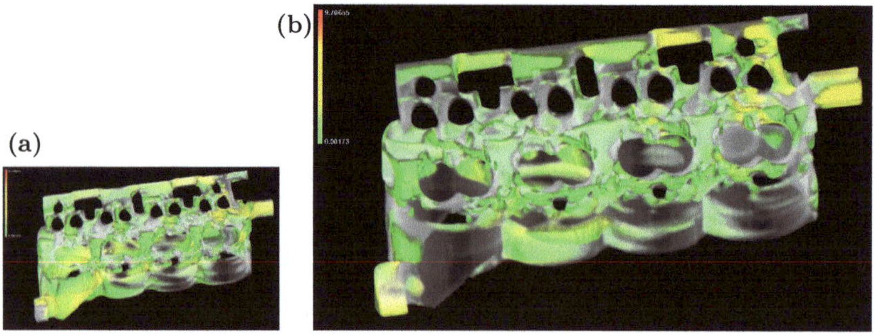

(a) (b)

Fig. 4 (p. 84). (**a**) The visualization of all regions of forward-transversal flow; color-mapping indicates velocity magnitude. (**b**) The result of selecting all regions of reverse-transversal flow, again the inverse of the left selection.

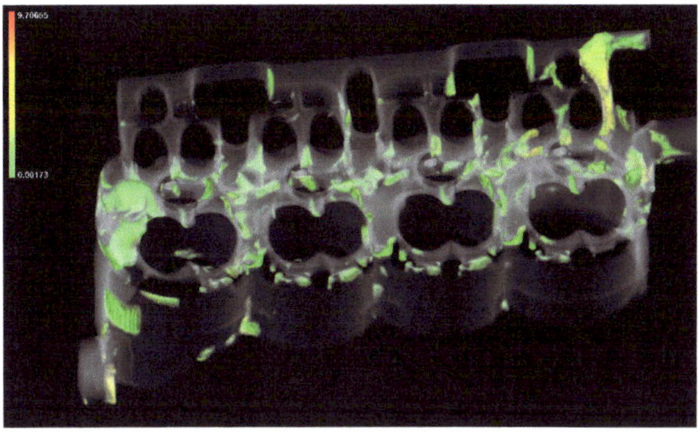

Fig. 5 (p. 85). The result of selecting all regions of reverse-longitudinal flow *and* regions of reverse-transversal flow.

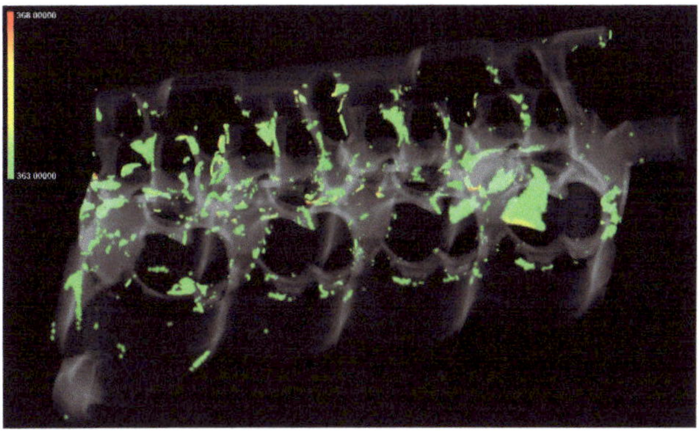

Fig. 6 (p. 86). A feature-based, focus+context visualization showing regions of near-stagnant flow, specified interactively.

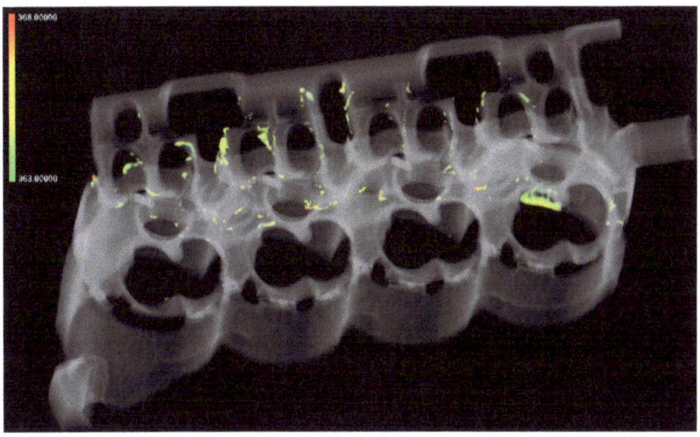

Fig. 7 (p. 87). Areas of temperature $t > 364\,^{\circ}K$ and velocity $|\mathbf{v}| < 0.1\,m/s$ are interactively-specified by the user and rendered in focus.

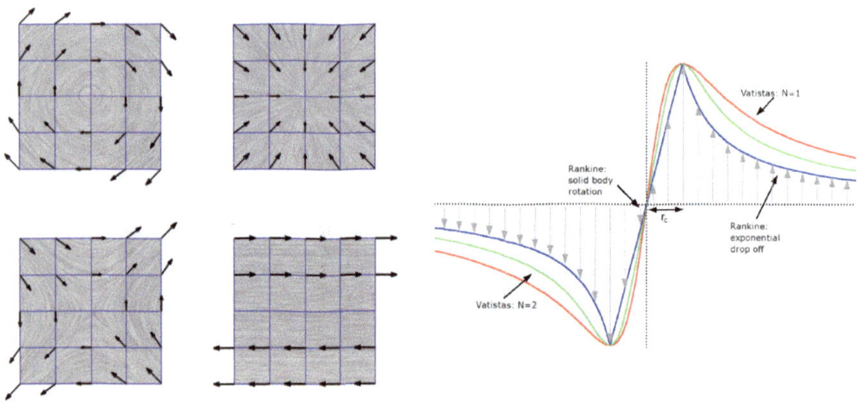

Fig. 2 (p. 94). Left: Rotation, convergence, saddle and shear, each visualized using hedgehogs and LIC. Right: The Vatistas vortex model. A pure, circular rotation is assumed. A cut through this rotation results in the described velocity profile.

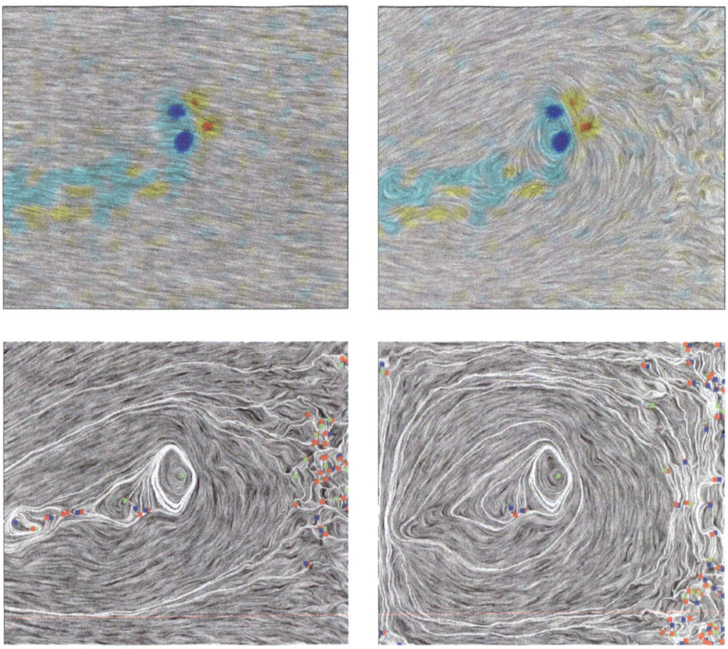

Fig. 3 (p. 98). One vector field of the HART II test measurements. Top left: LIC and vorticity of the original data set, dark blue: high negative vorticity, red: high positive vorticity. Top right: Vorticity and LIC of the data set after removing the average. Bottom left: LIC and topology after removing the average. Bottom right: LIC and topology of the region-specific flow of the dataset.

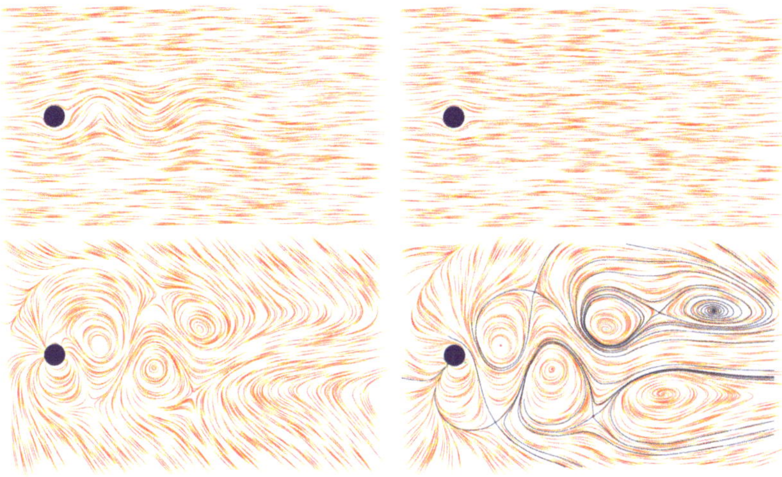

Fig. 4 (p. 99). Comparison of different fields obtained from a cylinder data set with a Kármán vortex street. Top left: Streamlines in the original flow. Only sinuous structures of the lines give hints on the vortices. Top right: Potential flow induced by the boundary. Bottom left: Three vortices revealed by removing the average flow. Bottom right: Subtracting the potential flow reveals all five vortices by use of topology.

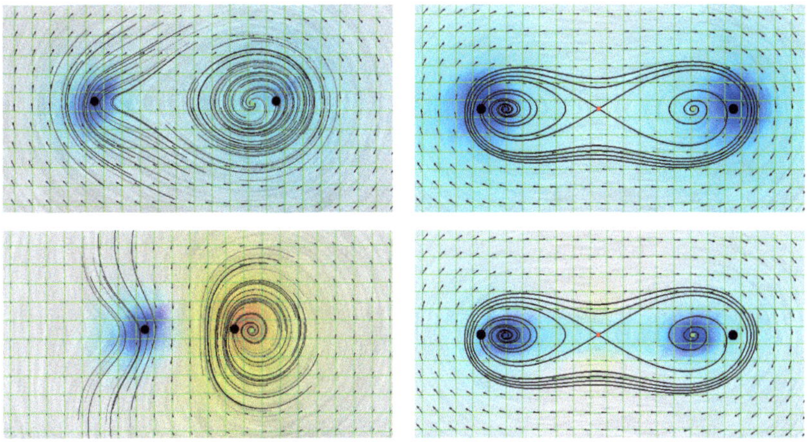

Fig. 5 (p. 101). Superposition and interaction of two Vatistas vortices. All images: The original vortex centers are displayed as black dots. Grid (green), hedgehogs (black arrows), color coding of similarity to a 3x3 rotational mask from high negative values (blue) to high positive values (red). Left: The stronger vortex hides the weaker vortex in streamline based visualizations. Topology only detects one center each (green dot), therefore some streamlines are added. Top right: Template matching detects the true vortex centers. Bottom right: Setting all velocity magnitudes in the field to one, and matching afterwards, yields results more similar to the topological features.

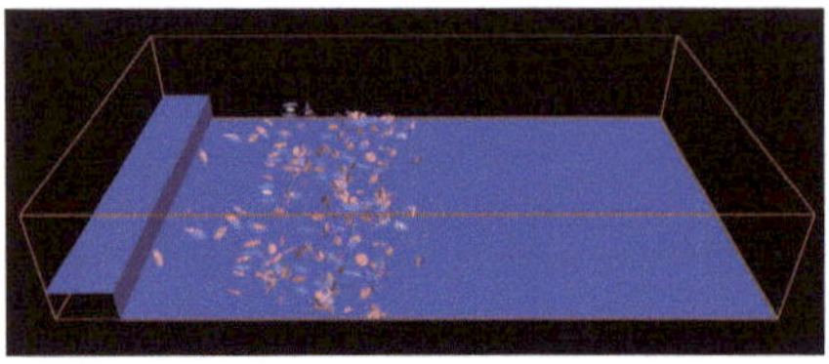

(a) All 452 critical points (all of them are saddles) are located in a rather small area of the whole domain.

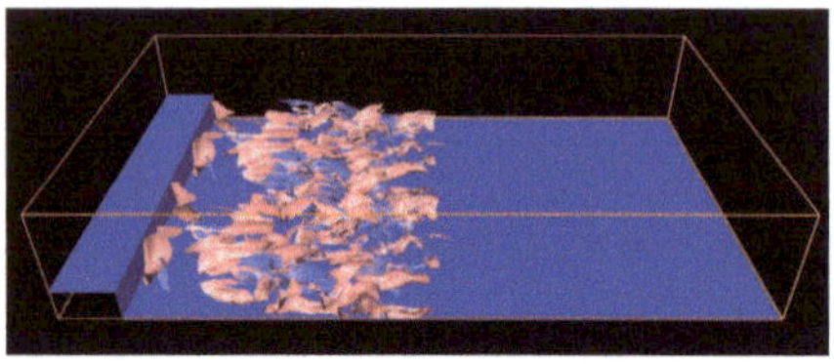

(b) 218 attracting and 234 repelling separations surfaces at a coarse resolution.

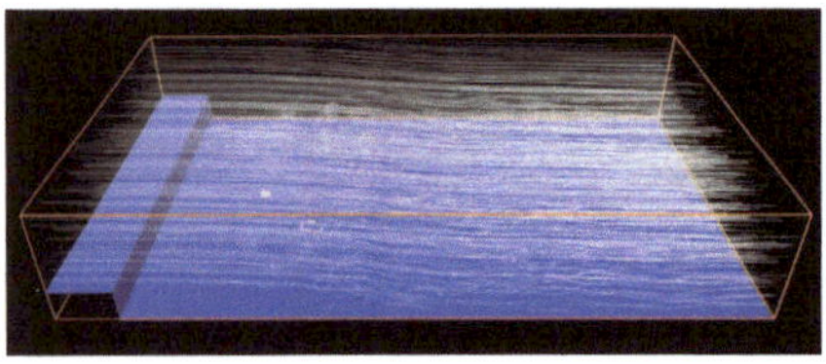

(c) 500 stream lines seeded homogeneous in the whole domain.

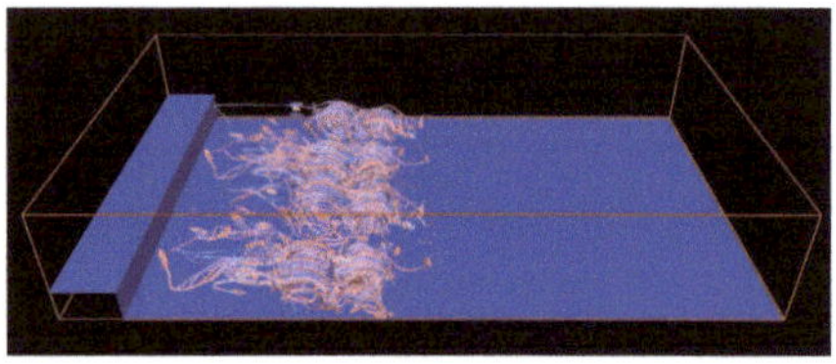

(d) 1023 saddle connectors have been extracted.

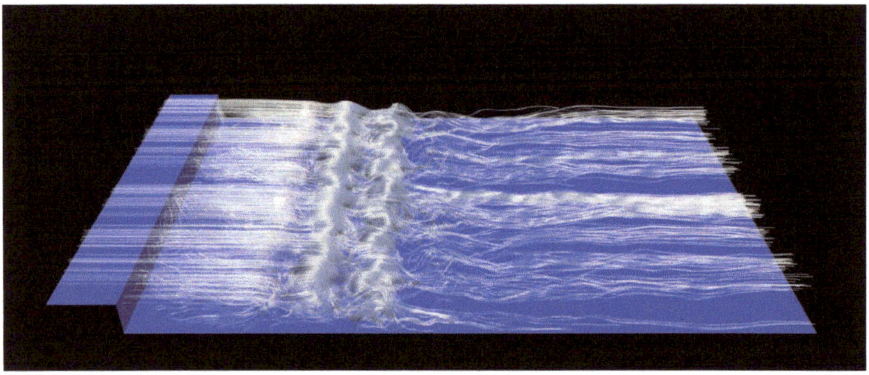

(e) 500 stream lines seeded near critical points.

Fig. 5 (p. 117). Flow around a backward-facing step.

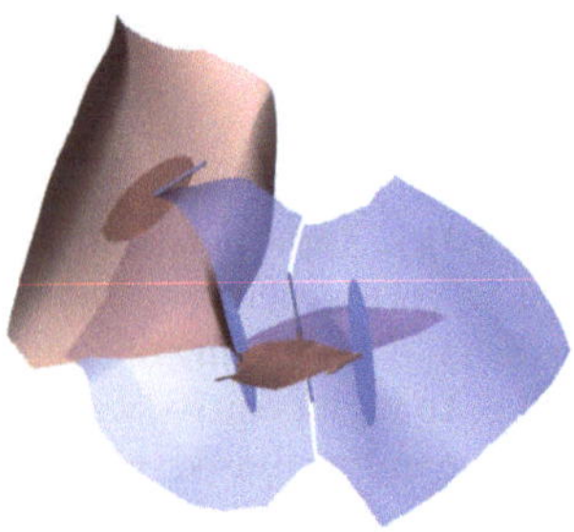

Fig. 6 (p. 117). Simple topological skeleton consisting of 4 saddles; the 6 resulting sectors of different flow behavior can hardly be distinguished.

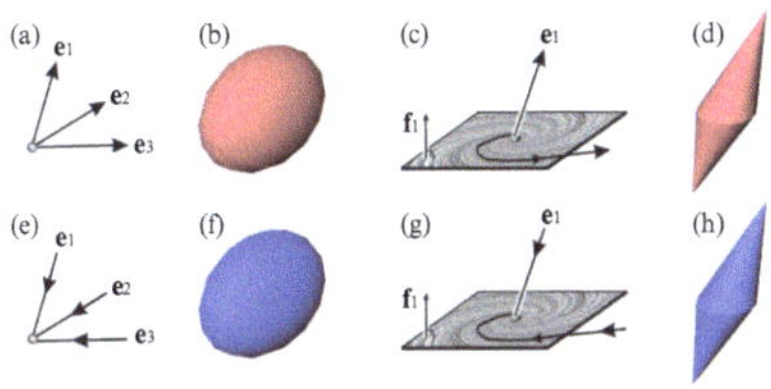

(a) Sources & sinks and their icons.

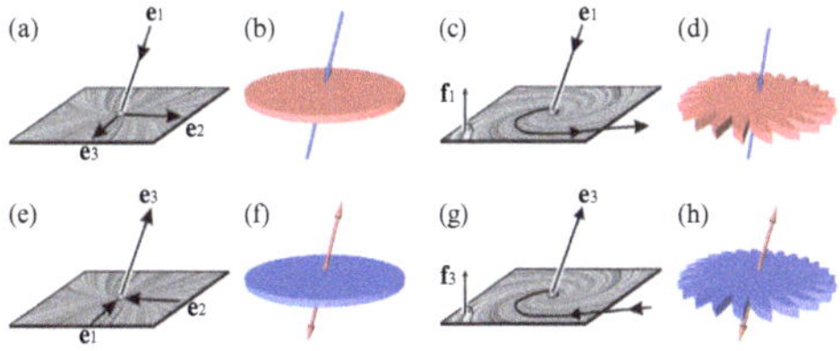

(b) Saddles and their icons.

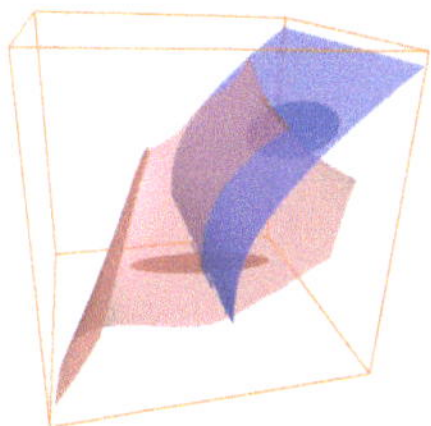

(c) Separation surfaces of the saddles.

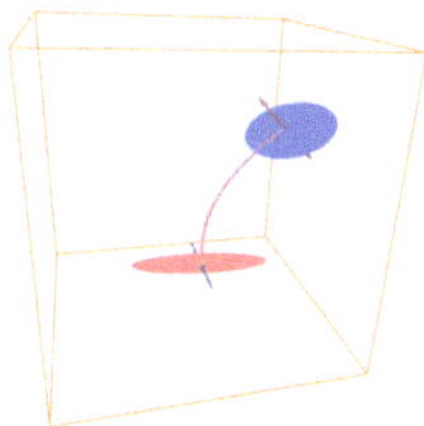

(d) Saddle connector.

Fig. 8 (p. 119). Critical points and definition of saddle connectors.

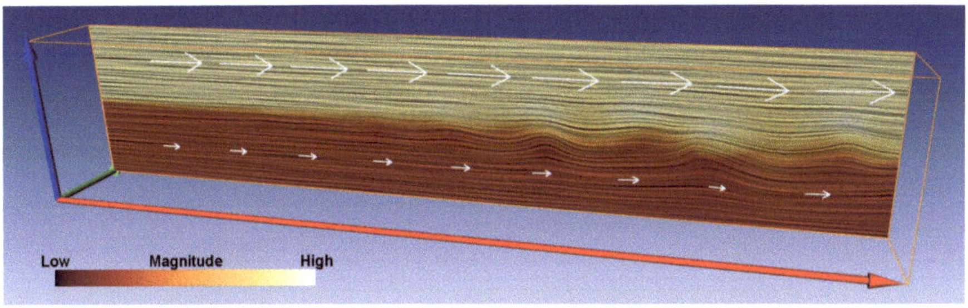

(a) Original frame of reference. No critical points are present.

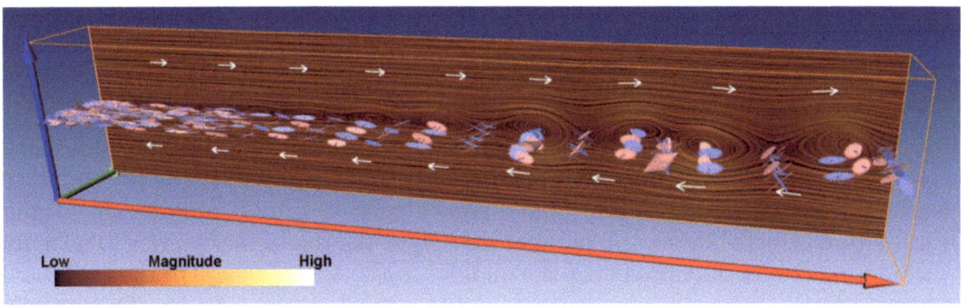

(b) Frame of reference chosen such that both layers have the same magnitude. 348 saddle points have been detected.

Fig. 7 (p. 118). Mixing layer.

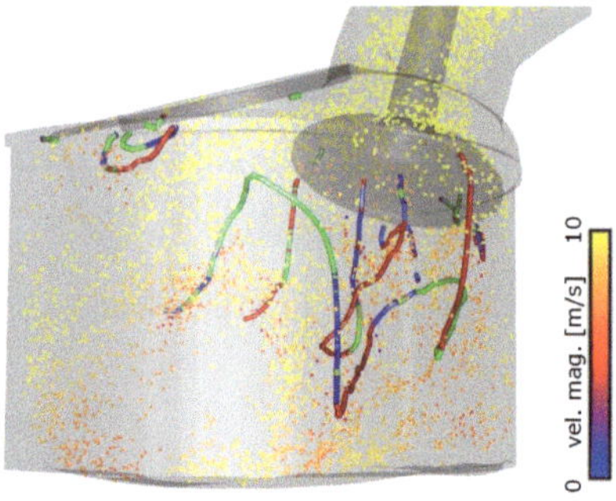

Fig. 2 (p. 126). A frame from a time-varying tumble motion visualization using a combination of particles and vortex cores. (lines, extracted by cutting-plane topology). Particle velocity magnitude is color-coded. Lines are color coded by the path type (saddles - red, sources - green, sinks - blue). Over time, some of the particles are captured in the vicinity of vortex cores resulting in lost energy (lower velocity) for the creation of the tumble pattern.

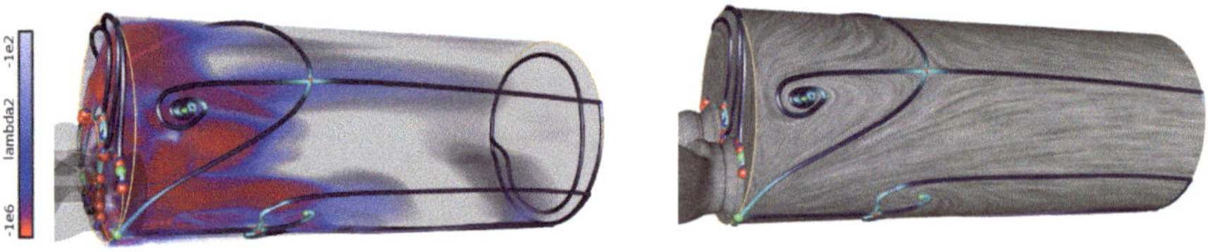

Fig. 3 (p. 127). Visualization of swirl motion using boundary topology. Critical points are colored by type (cf. Fig. 2), and separatrix color varies with separation/attachment behavior from dark blue (weak) to cyan (strong). Separatrices indicate the separation between neighboring vortices on the boundary. (Left) Combination with volume rendering with a transfer function of λ_2 only. On the bottom left of the cylinder, the recirculation zone causes a non-ideal off-center rotation, as visualized by topology. (Right) In combination with LIC.

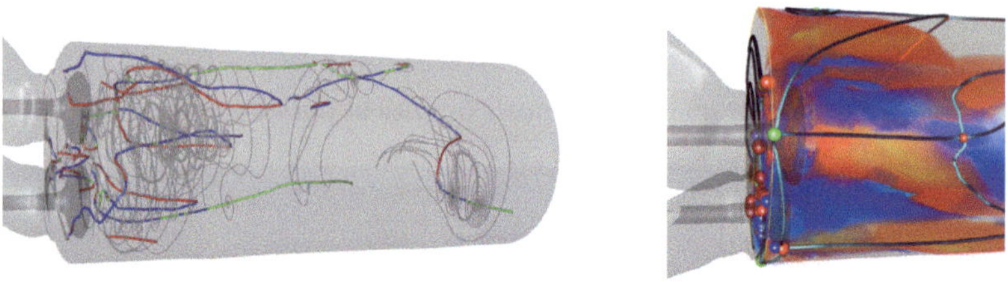

Fig. 4 (p. 128). *Left:* Cutting-plane topology applied to the diesel engine. Plane separatrices are colored gray. Despite the visual inexactness, swirl structures emerge clearly. Critical points paths are colored according to nature. It appears the overall swirl motion is fueled by several parallel vortices at the top of the cylinder. The main swirl motion core is disrupted near the middle. *Right:* Visualization of the rotational directions in the vortex system at the top of the diesel engine cylinder. The transfer function is identical to that of Fig. 6. The counter-rotating vortices appear in blue and in red depending on rotation direction. On the boundary, the topological analysis extracts and visualizes separation lines between individual vortices (color coding as in Figs. 2 and 3).

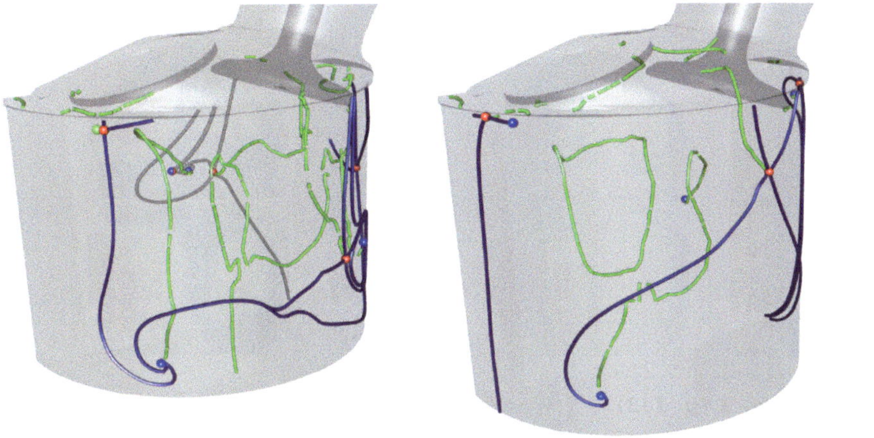

Fig. 7 (p. 132). Two frames from a time-varying visualization of tumble motion using a combination of cutting-plane topology (green lines) and boundary topology (blue lines). Where the critical point paths computed over the cutting-plane parameter range intersect the boundary, singularities appear there, too. This is a prime example of a hybrid approach being used to investigate the relationship between boundary and volume methods.

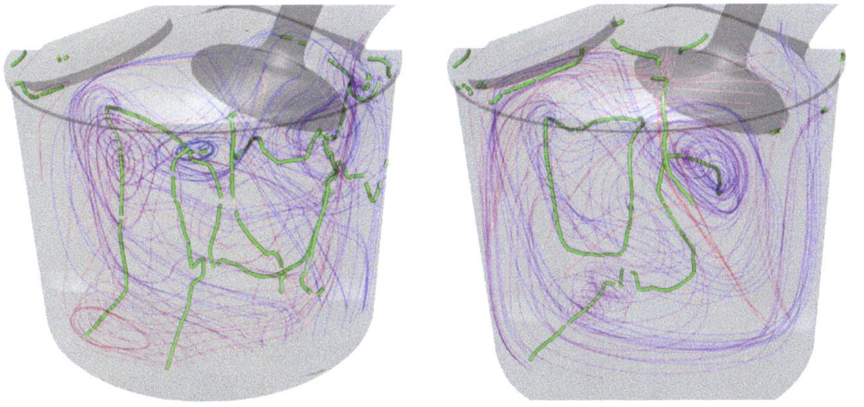

Fig. 5 (p. 129). Two frames from an animation of the tumble motion simulation. Cutting plane topology is applied to visualize flow field structures in the plane orthogonal to the tumble axis. Color of separatrices varies from blue to red on successive cutting planes. Tumble-like flow structures emerge clearly from the otherwise incoherent lines. The paths of critical points over the cutting plane continuum are displayed in green. In the last frame (right), the diagonal main tumble axis can be observed together with a large recirculation zone (closed path on the left).

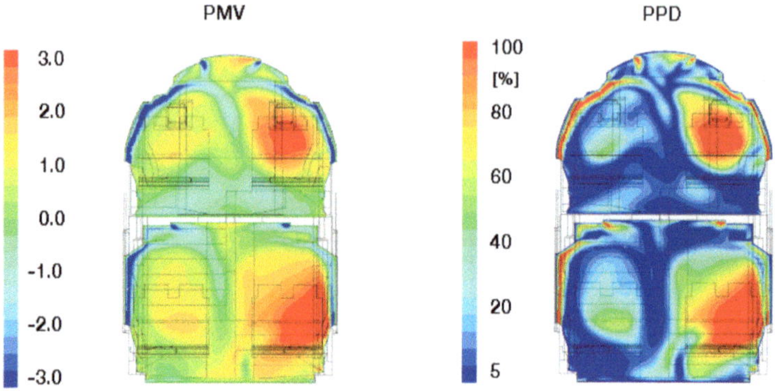

Fig. 4 (p. 142). *PMV* and *PPD* values for a section of the train.

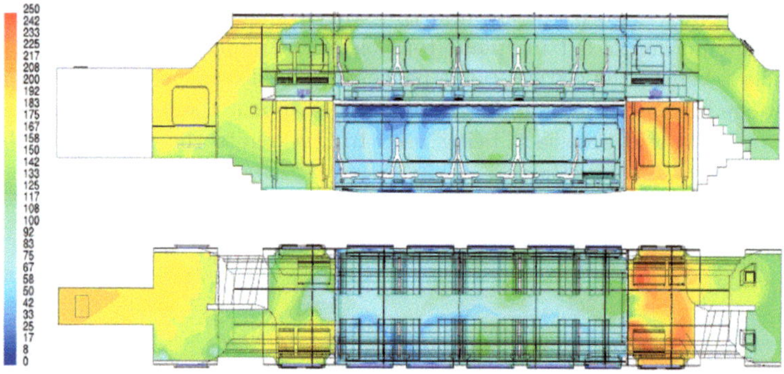

Fig. 5 (p. 142). *MAA* in [s] in the train.

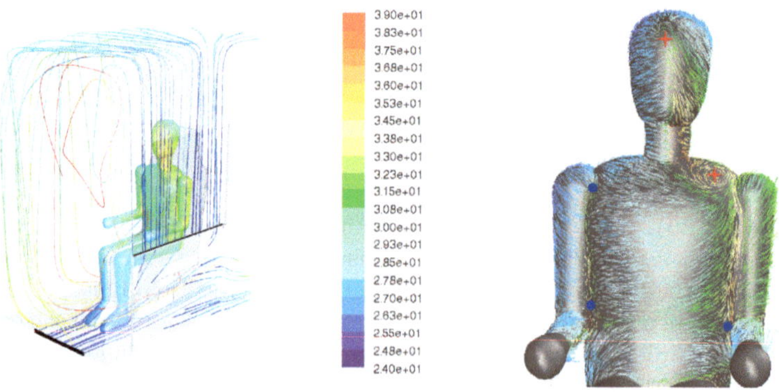

Fig. 6 (p. 143). The thermal mannequin in the compartment. Left: Visualization of streamlines, coloured with their residence time in [s]. Surface static temperature from 291K to 310K. Right: Texture based visualization of wall stream lines coloured with the local t_{eq} in $°C$ (colorplate on p. 216).

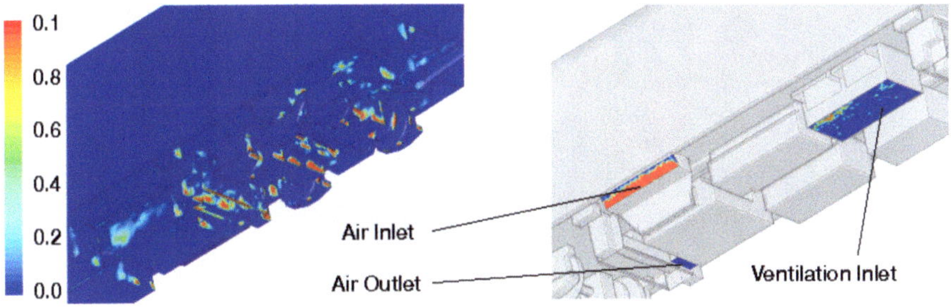

Fig. 9 (p. 147). Snow accretion in kg/m^{2}s on wall boundaries.

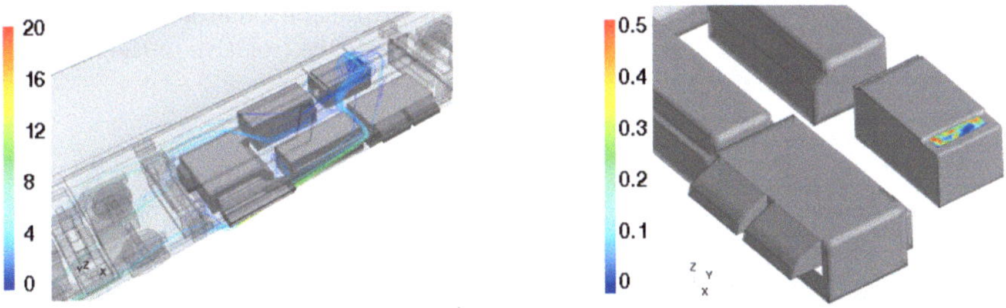

Fig. 10 (p. 147). Snow accretion in kg/m^{2}s on the compressor inlet and particle tracks coloured with the local particle velocity in m/s collected by the compressor.

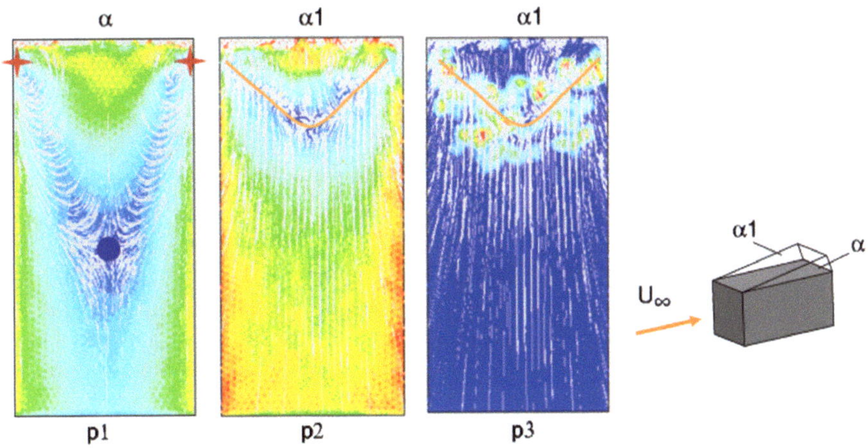

Fig. 11 (p. 148). Visualization of wall stream lines and accretion rate together with critical points for different angles of attack of the top surface of the box (colorplate on p. 217).

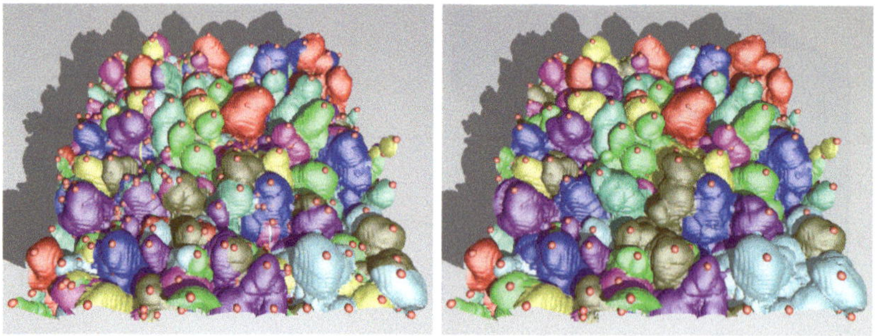

(a) (b)

Fig. 12 (p. 165). Segmentation of parts of the Mixing Fluids data sets into bubbles. Maxima are shown in red and each bubble is randomly assigned one of nine colors. (**a**) Initial segmentation; (**b**) Segmentation at 0.2% persistence (relative to the maximal range in function value).

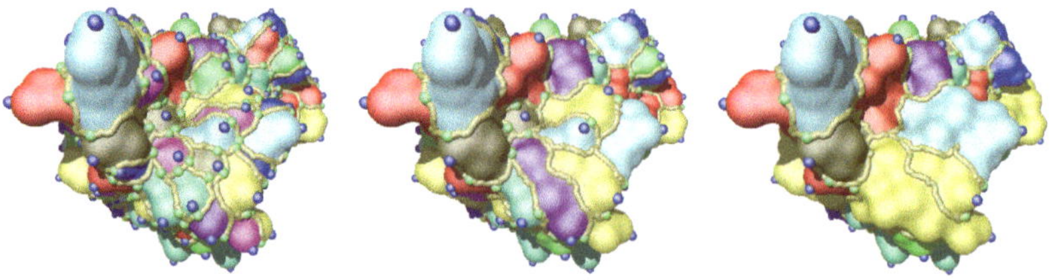

(a) (b) (c)

Fig. 13 (p. 166). Segmentation of the atomic density function on a molecule. Minima are shown in blue and ascending paths in gold. Segmentation into 198 (**a**); 100 (**b**); and 50 (**c**) protrusions.

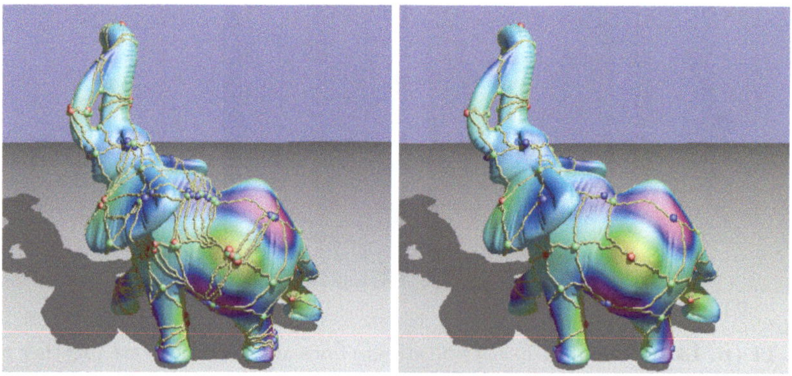

(a) (b)

Fig. 14 (p. 167). Creating a basemesh from Laplacian eigenfunctions. (**a**) Initial MS complex of the 60th eigenfunction of the elephant showing typical noise due to discretization and the iterative eigensolver. (**b**) The MS complex of (a) simplified to a persistence of 0.5%. All noise has been removed and the MS complex now forms a high quality all-quadrilateral basemesh.

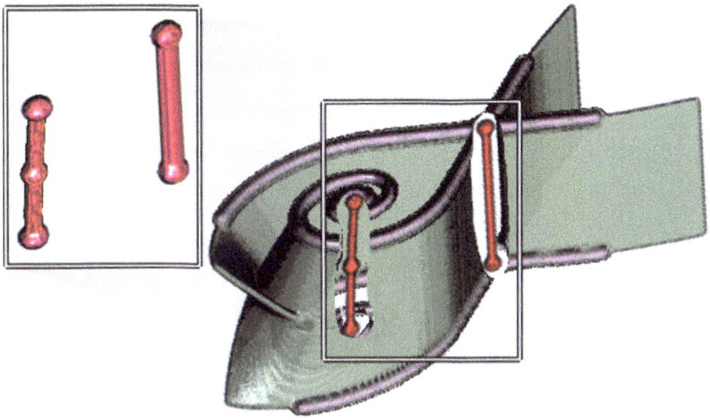

Fig. 1 (p. 172). Periodic Blue Sky Bifurcation. The polygonal model is courtesy of Tino Weinkauf and Holger Theisel.

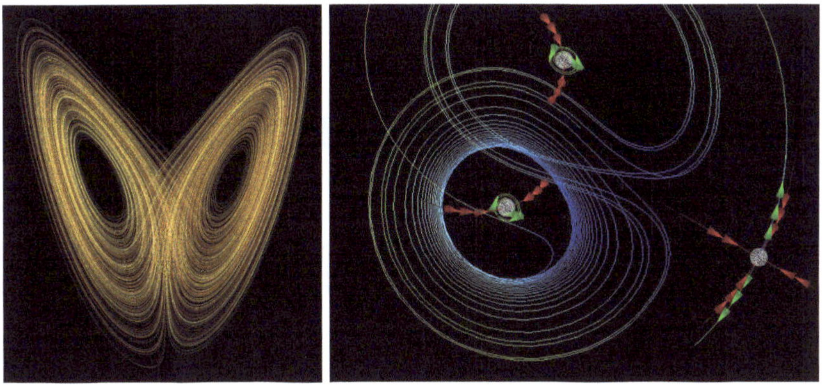

(*a*) (*b*)

Fig. 2 (p. 174). Lorenz attractor: (a) direct representation through a streamline and (b) descriptive visualization of the system with emphasis of critical points [4].

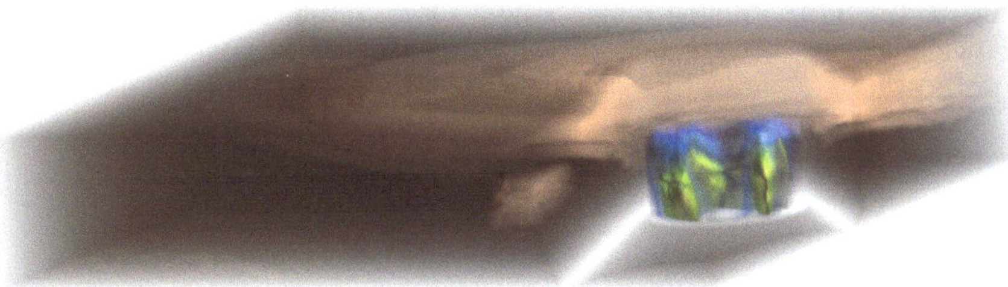

Fig. 5 (p. 178). Cut-away visualization of a multidimensional data of hurricane Isabel.

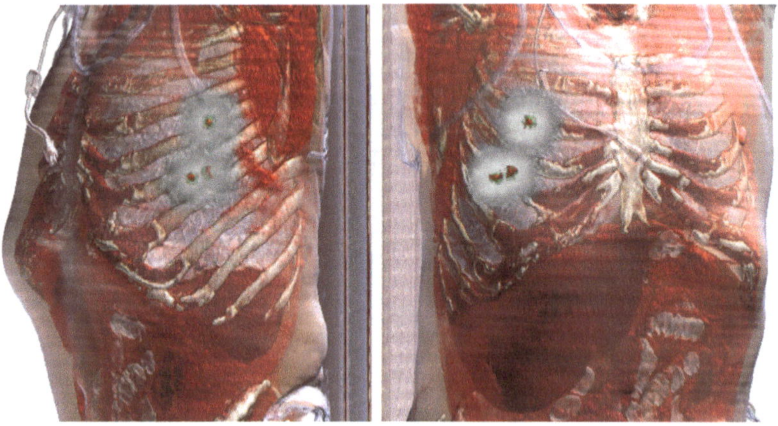

Fig. 4 (p. 177). Topology-driven visualizations of suspicious regions (possibly early-stage lung cancer nodules) within a thorax CT data. The suspicious regions are highlighted in red enclosed by a green halo.

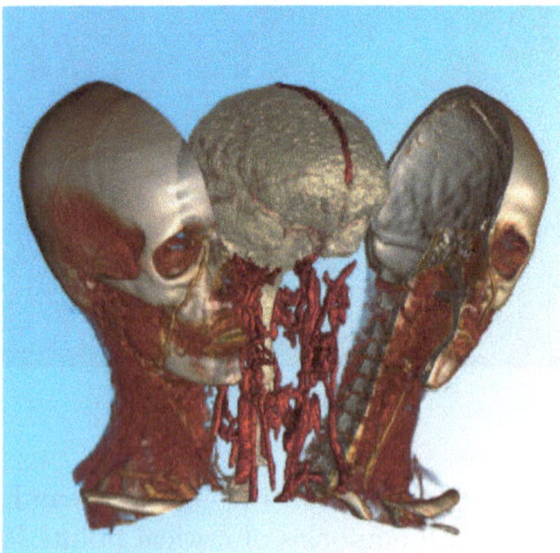

Fig. 6 (p. 179). Multi-volume visualization enabling exploded views through topology change of the underlying data [1].

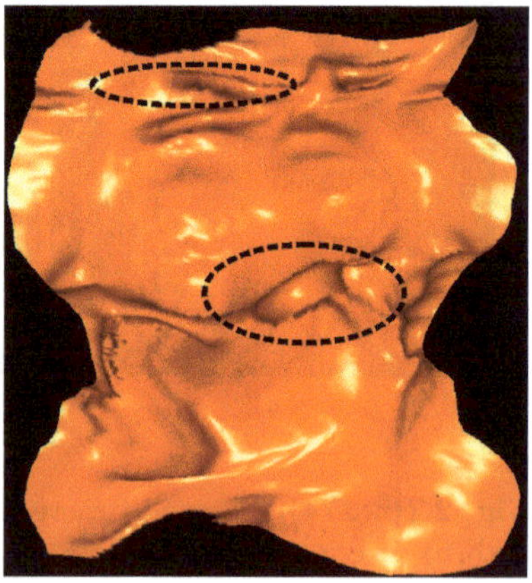

Fig. 7 (p. 180). Virtual colon unfolding using a topology change in the visualization [10].

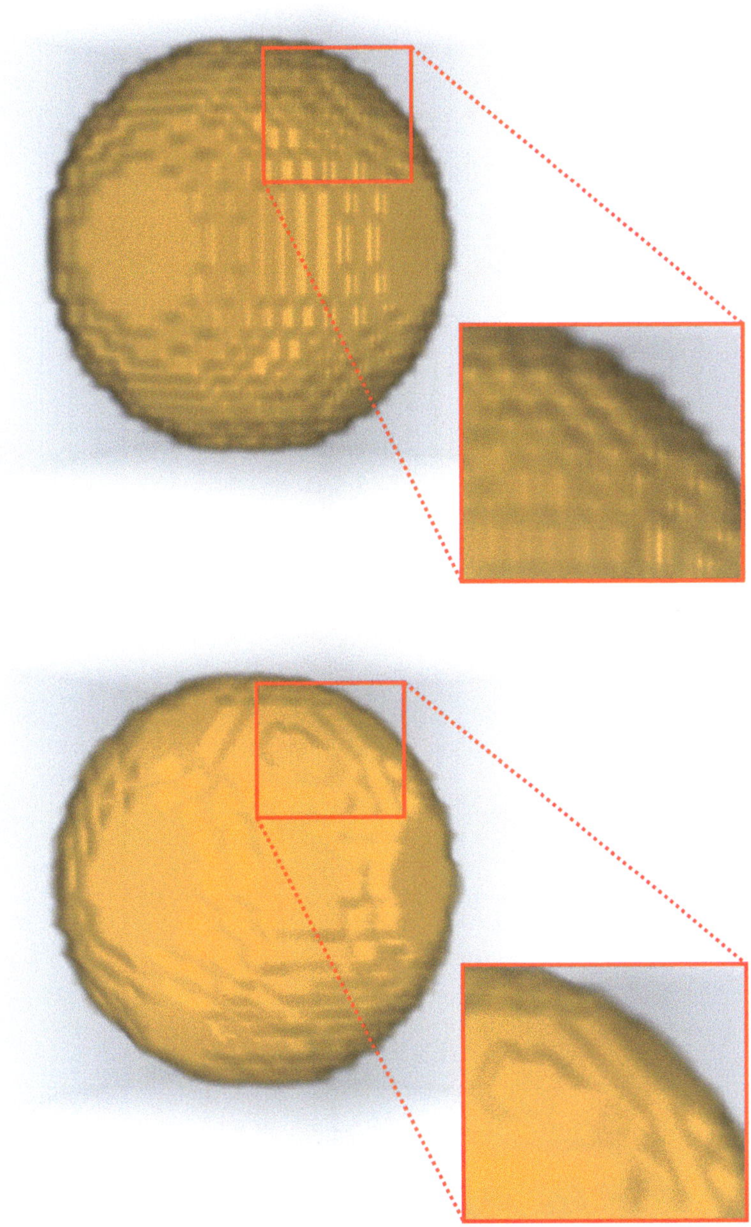

Fig. 9 (p. 195). Results from trilinear interpolation (top), and from the generalized Lawson's optimization DDT approach with the variance-based weight function and bistellar flips (bottom).

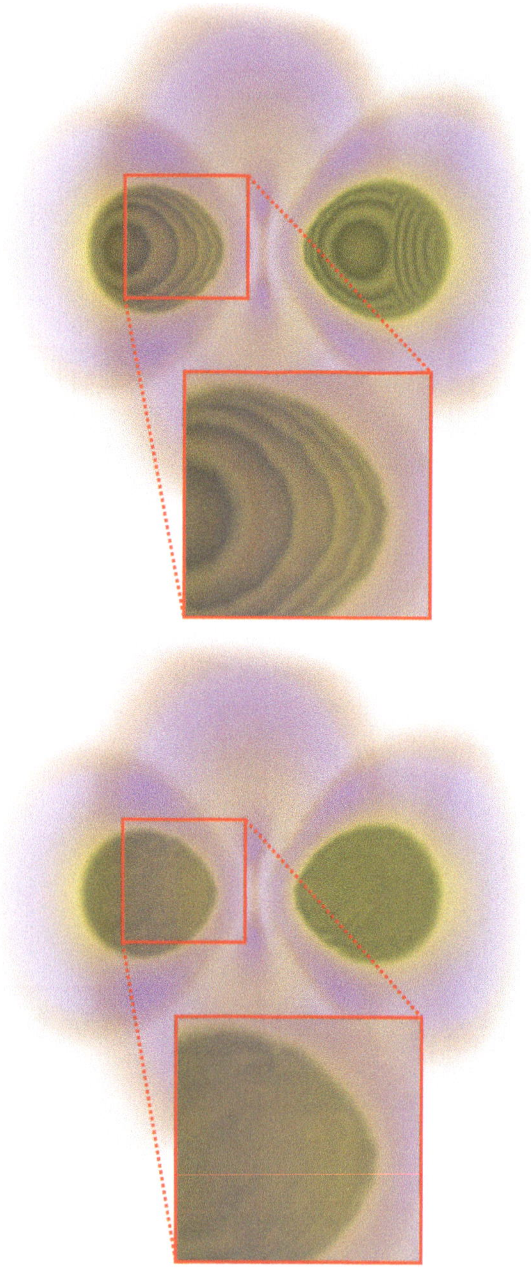

Fig. 10 (p. 196). Hydrogen dataset: The top image shows the result from trilinear interpolation suffering from noticeable artifacts. The bottom one shows improved results achieved by DDT-based reconstruction with bistellar flips and the variance-based weight function.

MIX
Papier aus verantwortungsvollen Quellen
Paper from responsible sources
FSC® C105338

If you have any concerns about our products, you can contact us on
ProductSafety@springernature.com

In case Publisher is established outside the EU, the EU authorized representative is:
Springer Nature Customer Service Center GmbH
Europaplatz 3, 69115 Heidelberg, Germany

Printed by Libri Plureos GmbH
in Hamburg, Germany